D0910149

Science and International Affairs Series
Melvyn B. Nathanson, Editor

CRISIS CONTAINED

The Department of Energy at
THREE MILE ISLAND

By
Philip L. Cantelon *and*
Robert C. Williams

Southern Illinois University Press
Carbondale and Edwardsville

TULSA JUNIOR COLLEGE
Learning Resources Center
Campus

Foreword and Introduction copyright © 1982 by the Board of
    Trustees, Southern Illinois University
All rights reserved
Printed in the United States of America
Production supervised by John DeBacher

Library of Congress Cataloging in Publication Data
Cantelon, Philip L. (Philip Louis), 1940–
    Crisis contained.

    (Science and international affairs series)
    Bibliography: p.
    Includes index.
    1. Three Mile Island Nuclear Power Plant (Pa.
2. Atomic power plants—Pennsylvania—Accidents.
3. United States. Dept. of Energy.   I. Williams,
Robert Chadwell, 1938–   II. Title.   III. Series.
TK1345.H37C36   363.1'79   81-21413
ISBN 0-8093-1079-1   AACR2

CONTENTS

ILLUSTRATIONS

FOREWORD

Melvyn B. Nathanson

Public policy is becoming increasingly technological. Arms control, nuclear nonproliferation, toxic waste disposal, energy, and environmental protection are problems that require both scientific and political wisdom. The most controversial technological issue, however, is nuclear energy and the safety of commercial nuclear reactors. In the debate about nuclear safety, it is necessary to know how government and industry would react to an accident in a nuclear power plant. *Crisis Contained* is an important contribution to understanding the governmental response to a nuclear emergency.

Crisis Contained is a remarkable book. The Department of Energy (DOE) commissioned Philip L. Cantelon and Robert C. Williams to write the official history of DOE involvement in the accident at Three Mile Island on March 28, 1979. The greatest danger to the public was the possible release of radioactivity into the atmosphere. Under the Interagency Radiological Assistance Plan (IRAP), the Department of Energy was responsible for monitoring radiation in a nuclear emergency. Beginning a few hours after the accident, DOE personnel began measuring the radioactivity released from the damaged reactor. Very little radioactive iodine and other harmful forms of radioactivity were actually released, and people living near Three Mile Island were in no danger. Nonetheless, panic spread around the reactor site and throughout the country. Public officials in the Nuclear Regulatory Commission, the Environmental Protection Agency, and the White House were not aware of the data being collected by IRAP scientists, and talked ominously of a meltdown and evacuation. After the accident, Three Mile Island was invaded by politicians and representatives of many different state and federal agencies and public and private interest groups. Pennsylvania Governor Richard Thornburgh complained that "there are a number of conflicting versions of *every event* that seems to occur." No one was in charge.

Crisis Contained presents a daily, sometimes hourly account of DOE activities at Three Mile Island. It is a study in bureaucratic process: What official in what section of what department was responsible for what part of the work to protect the public and to shut down the failed nuclear reactor. The book describes how DOE responded quickly and efficiently to the accident in accordance with its previously prepared plan for a nuclear emergency. The book also shows how little the various public agencies present at Three Mile Island knew of each other's activities, capabilities, and responsibilities.

The chief casualty at Three Mile Island was public trust in the government and the nuclear industry, which had for years exaggerated the safety of commercial nuclear reactors. In 1976, an official government study of nuclear safety predicted that even if 1,000 nuclear reactors were simultaneously in operation, a major accident, a meltdown of a reactor core, would occur only once in 10,000 years. At Three Mile Island, a minor mechanical failure was made worse by a series of wrong decisions by the men in the control room. This almost led to the meltdown that the Atomic Energy Commission, the Nuclear Regulatory Commission, and the nuclear industry had claimed was impossible, and it changed the public vision of the future of nuclear energy in the United States. To understand the events at Three Mile Island, it is important to read *Crisis Contained*.

INTRODUCTION

Three years after America's worst nuclear power accident, it is clear that Three Mile Island was not so much a technological event as a human and historical one. It was a failure not of technology but of technocracy, the rule by experts based on criteria of rationality, efficiency, productivity, and profit. It was not only a mechanical breakdown but a series of human choices that crippled a nuclear reactor and threatened injury and death to the public. What escaped at Three Mile Island was not only radiation, but, more importantly for the nuclear power industry, public confidence in technology and technocracy.

Today, the plant remains crippled, and the radiological, economic, political, legal, and psychological effects of the accident will be felt and scrutinized for years. Three Mile Island has joined Skylab and Titan as a symbol of technological failure. Yet the serious extent of the accident was caused by human error: technocrats blundered, lost control of technology, and, refusing to admit it, gave confusing, inconsistent, and jargon-laden explanations. (How many Americans knew what was meant by a "hot leg," a "candy cane," or a "primary loop"?) As a result of the failure of government officials and scientists to communicate to the public, the line between real and imagined risk became blurred: citizens were no less traumatized because the event happened to them emotionally. Risk perceived is risk endured.

Those who felt their lives threatened do not accept the reality: a minor loss-of-coolant accident that caused them less radiation than their dentist and was less likely to kill them than their automobile. They feel they have lived through hell. They have come to distrust statements from an industry that calls a bomb a "device" and an accident an "incident." They remain confused, afraid, and angry.

In trying to reconstruct and explain the human past, history should be part science and part art. Like a scientist, the historian deals in evidence, inference, and logical argument; like an artist, he creates aesthetic order out of the chaos of experience. The historian must narrate a reasonably true story of events based on documentary evidence, but he must also understand through empathy the actions, motives, ideas, hopes, and fears of human beings at other times and places.

History, science, technology, and the humanities are united in the story of America's worst civilian nuclear power accident. In a superficial and ironic sense, one common denominator is Faust. Craig Faust, a young man trained in Admiral Hyman G. Rickover's naval reactor school, was on duty at 4:00 A.M. on March 28, 1979, when Unit 2 at Three Mile Island began to malfunction. The Faust of literature, of course, is a sixteenth-century learned man who makes a fateful pact with the Devil to understand the innermost secrets of nature in order to dominate it.

In another sense, nuclear energy promised a new utopia. Nuclear power, according to its early proponents, was a safe, renewable, cheap, even limitless source of energy.

The related tradition in Western thought is Sir Thomas More's *Utopia* (1517) which took its title from a pun on Greek words for "the good place" and "nowhere." Like subsequent utopias, More's was an imaginary, idealized community situated on an island (a good place for imagining a heaven on earth). Perhaps the best example of a scientific utopia is Francis Bacon's *New Atlantis*, which describes a seventeenth-century fictional island society where the scientific elite works miracles through experiment, exchange of ideas, and technology.

But utopia has its price: we give up our freedom to those who promise to plan our happiness. Technocracy supersedes democracy. As the Russian writer Aleksander Solzhenitsyn never tires of reminding us, the utopian aspirations of Western secular humanism since the Renaissance have repressed or ignored essentially religious values, which he finds more apparent in the suffering islands of the Gulag Archipelago than the happy islands of Western utopias.

In the context of Three Mile Island, utopia was a dream of secular abundance through nuclear power. The opposite of that utopia was a "nuclear nightmare" that resembled Dante's fourteenth-century literary vision of hell in *The Divine Comedy*—the island of Mount Purgatory surrounded by seven circles corresponding to the seven deadly sins. Here man is not godlike but sinful. At Three Mile Island, hell took the form of concentric circles around the plant, circles which plotted the plume of radioactive gases and the contingencies of evacuating hundreds of thousands of people from the area. Three Mile Island, then, may be thought of as a symbol of nuclear utopia encircled by the imagination of nuclear catastrophe.

In writing the history of the Department of Energy at Three Mile Island, we focused on the emergency response efforts made by local, state, and federal officials. The main problem was to monitor offsite radiation and take measures to protect the public from it, or from some more catastrophic possibility. Emergency response was the primary function of the Environmental Safety branch of the Department of Energy; by Wednesday afternoon Department of Energy teams were monitoring ground and aerial radiation around the plant and had established a command post at nearby Capital City Airport. By Friday afternoon, a second wave of politicians and other government officials descended on the area on White House orders to bring more expertise to the aid of the plant.

Department of Energy scientists soon found that reality was hard to establish. Almost no information was available from Metropolitan Edison about the plant itself. Regular helicopter flights were made around the clock in hopes of catching releases, but these releases were often unannounced, or announced and then cancelled. Emergency teams brought very sensitive instruments used in looking for tiny radioactive sources, such as the fragments of the Soviet nuclear-powered satellite lost over Canada in 1978, and were astounded when they went off scale: they knew that radiation was higher than they expected, but not how much higher. Other teams, following up early (erroneous) readings by Met Ed, came looking specifically for iodine-131, an isotope that does severe damage to the thyroid and liver; finding no iodine, they felt that there really was no emergency at all, and asked to be sent home.

In general, scientists did not know what to expect because communications were so poor. They had no data from the plant itself, and did not know what was happening to the information on the very low radiation they and the Pennsylvania Bureau of Radiation Protection were regularly reporting to the Nuclear Regulatory Commission; these readings were lower than those monitored in the same area after a 1976 Chinese nuclear bomb test. They could not understand why they were seeing government officials on television speculating about evacuation, explosion, and meltdown. The public excitement over Three Mile Island did not make sense to scientists in terms of the levels of radiation they were monitoring, and they tended to ascribe this to politics and the media.

At Three Mile Island thousands of citizens had to choose whether or not to leave their homes for safety, and a number of politicians and government officials had to decide whether or not to advise them to do so. Evacuation of people from areas endangered by floodwaters is common in southern Pennsylvania, but there were no plans for evacuation in connection with a nuclear power plant accident. Evacuation was plotted on a series of concentric circles with radii of five, ten, or twenty miles. Like so much else at Three Mile Island, evacuation turned out to be imaginary: the major step was to publish in Monday's *Harrisburg Patriot* maps of how to get out of the area.

The reality was not evacuation, which implies orderly planned action by government, but exodus and flight. Over the weekend of March 30 and April 1 the people of Three Mile Island voted with their feet. They did not wait for an evacuation order that never came (aside from an advisory by Governor Thornburgh that preschool children and pregnant women leave the immediate area), but got in their cars and left to spend the weekend with friends or relatives, some as far away as St. Louis.

Politically, Three Mile Island had happened in the public imagination well before it happened in reality. Technocracy created nuclear power; democracy rarely voted for it. Throughout the 1970's a public fear of science and technology had been reflected in popular hostility, tight environmental regulations, and reduced federal funding. Shortly before the accident, the Nuclear Regulatory Commission had withdrawn its approval of the conclusions of an earlier report by Professor Norman Rasmussen, which predicted that serious nuclear accident, such as a meltdown, was extremely improbable. Citizen groups, such as the Clamshell and Crawdad Alliances, attempted to halt further plant construction and licensing. Despite mounting costs of oil, many orders for nuclear plants were declining or cancelled well before Three Mile Island. By 1979 the future seemed to belong to the snail darter, not the Clinch River Breeder Reactor.

Fictional accounts of nuclear nightmares also sold well. *The Prometheus Crisis* (1975) urged us to "imagine a hundred thousand towering infernos— with effects lasting for generations to come." The novel *A Short Life* (1978) warned that "nuclear disaster is no longer a threat—it is a reality that is happening today as accidents in supposedly 'safe' power plants release clouds of radiation into our atmosphere" and *The China Syndrome*, which appeared as a film coincidental with the accident, promised that a nuclear accident "means something very hot, very explosive. Something so big it could blow Southern California sky high!" Or an area the size of Pennsylvania.

Given popular expectations of nuclear disaster, accurate communication of the risks to public health and safety were crucial at Three Mile Island. For three days the utility acted as if nothing was wrong. Only on Friday, when the Nuclear Regulatory Commission learned of Wednesday's hydrogen explosion and the existence of a noncondensible bubble in the reactor, did the federal government act to improve communications by sending Harold Denton and a team of experts to Three Mile Island. But even then the experts disagreed: scientists at the Idaho National Engineering Laboratory knew by Saturday that the bubble could not explode because of insufficient oxygen; they reported this to the Nuclear Regulatory Commission. Yet when President Carter arrived in Middletown early Sunday afternoon, the two chief Nuclear Regulatory Commission reactor experts still disagreed on the explosive potential of the bubble. In the end the Nuclear Regulatory Commission never did admit that its calculations of a possible explosion were simply erroneous. In general, the problem of the hydrogen bubble illustrated a more common one—scientific data were not being considered before political decisions were being made, or before statements were released.

Not that it would have mattered. When a Nuclear Regulatory Commission official, Dudley Thompson, used the term "meltdown" at a Friday afternoon news conference, no one noticed the laconic 6:30 P.M. Nuclear Regulatory Commission bulletin that said there was now "no immediate danger of a core melt." What people heard and saw was Walter Cronkite's "nuclear nightmare" coverage on the evening news. One Nuclear Regulatory Commission official speculating out loud on the remote possibility of a meltdown thus was amplified in the public mind. Only with Harold Denton's subsequent press briefings did the media and the public get what should have been available from the outset—a single and credible scientific expert able to communicate reality.

In fact, all weekend the scientific data indicated low radiation levels and minimal danger of another explosion. But these data were not effectively communicated. Too many decisions were made on the periphery, in Washington and Bethesda, and not at the center of events. Scientific data did not inform political decisions, and the media generated alarm based on contradictory and confusing government statements, often in a language incomprehensible to the layman.

Imagination and unconscious fears were far more important than any accurate perception of risk. "We all live in Harrisburg," the slogan went, not only because of geographic proximity or television, but because of the shared dread of radiological death in the nuclear age. Americans have not forgotten that nuclear power originated in weapons: the bomb and the submarine. Behind the nightmare of Three Mile Island may well lie guilt and the fear that we too may one day experience our own Hiroshima and Nagasaki. At Alamogordo in 1945 after the first atomic bomb test, J. Robert Oppenheimer recalled the words of the Hindu epic poem the *Bhagavad-Gita*: "I am become death, the shatterer of worlds."

Reports of psychological stress suggest that the main legacy of Three Mile Island may have been the survivor's sense that he has narrowly escaped death and life continues to be out of control. To many people a nuclear reactor is simply a silent bomb, regardless of constant statements that it cannot possibly explode; the unseen threat of radiation adds to this feeling. No wonder children living in the area have had the recurrent nightmares about death, and psychiatric patients have aged visibly under the strain. A grandmother says she now lives with a sense of doom; a physician reports emotional trauma, rather than physical effects; every stillborn litter of kittens or aborted sheep brings back old memories and associations; a housewife says, "It's like we don't have control over our lives." Of all the residents near Three Mile Island, only the Amish, devout believers who avoid contact with civilization, appear to have emerged relatively unscathed. Without television or newspapers, they learned of the accident more than a week after it happened; when they found out, they simply ascribed it to God's will and man's meddling with nature. The Amish still have what technocracy and utopia lack—a sense of man's original sin and essential imperfection.

Three Mile Island, then, should be understood as an event of historical significance not only because of what actually happened, but because of what people thought was happening or feared might happen. Time and again people made choices based on inappropriate conceptions of reality. Technology, technocracy, containment, emergency response, evacuation, politics, and mortality—these are not the seven deadly sins of Dante's hell, but varieties of human involvement in history at Three Mile Island. They remind us of the persistence of human choice in an age of science, technology, and specialization: we cannot blame technology for human error.

More broadly, Three Mile Island shows us that the human implications of science are too important to be left to the scientists alone. Technocracy must be ultimately responsible to democracy, especially when technology threatens public health and safety. Technical expertise is crucial to our civilization. But it is not appropriate if it loses control, if it cannot explain itself clearly in layman's terms, or if it threatens public health and safety. Three Mile Island was a human drama and trauma that is still with us; it is contemporary history here and now, not there and then. We still do not know when the accident will truly be over. For a moment science lost control; the perspective of the historian can help us understand why.

☆ ☆ ☆ ☆ ☆ ☆ ☆ ☆ ☆ ☆ ☆ ☆ ☆ ☆ ☆

This history is nearly contemporary with the events it describes. Begun in the summer of 1979, shortly after the accident, it was completed by June 1980 and issued as a government report that December. Because it is contemporary history, it is able to draw on the memories of participants, but it is not as comprehensive or complete as a full-scale history of the accident might be. Indeed, one might argue that until the professional historian emerged in the nineteenth century, much of history was contemporary with the events it described. Working under a government contract and operating within prescribed time constraints, we could not enjoy the luxury of the broad view and the long perspective. It may be years before a full-scale history of Three Mile Island appears, based on records presently unavailable, on additional special studies, and on the outcome of America's debate over the future of nuclear power.

Much of what passes for contemporary history today consists of myths and/or memories shaped by the instant analysis of television reporters, or the writings of newspapers and magazine journalists. Historians are rarely asked or expected to write about contemporary events, especially by government agencies interested in more immediate lessons learned for future policy making. The opportunity to write a history of the Department of Energy at Three Mile Island thus has been useful in two ways: it has brought the skills and perspectives of the historian to bear on a problem of contemporary public policy; and it has related a story of a government agency untold in the media. Telling the story of what happened to one federal agency during America's worst nuclear power plant accident thus provides a useful complement to the quick reporting of the journalist and the *ex post facto* recommendations of the investigator or policy maker.

Historical perspective is essential and unless historians evaluate the significance of the immediate, as well as the more remote past, history becomes little more than a list of events. Department of Energy personnel were not merely present at the Three Mile Island accident. They were significantly involved from well before the first day until long after public attention had shifted to other matters. More than one hundred scientists and technicians funded by the Department of Energy, people with long and intensive training in radiation safety, conducted environmental monitoring of radiation, predicted meteorological and radiological conditions, tested laboratory samples of contaminated materials, arranged for the possible evacuation of Department of Energy and Nuclear Regulatory Commission personnel, and supplied the Nuclear Regulatory Commission itself with everything from monitoring data, maps, and lead bricks to photographic services. A thorough understanding of the Department of Energy's role at Three Mile Island is thus significant in contributing to our understanding of the mechanism and effectiveness of federal responses to nuclear emergencies in general.

Unlike most historical events, the Department of Energy's role at Three Mile Island included prior planning. Historians have traditionally concerned themselves with causal explanations of unplanned events: why did the French Revolution occur in 1789? Why did World War I break out in August 1914? In this case, the historian knows why the Department of Energy was at Three Mile Island: because it was responding according to a long-standing but little known Interagency Radiological Assistance Plan. In this situation a different set of questions is in order. What was expected to happen in the event of a nuclear accident? Did events occur as anticipated when an accident actually did happen? Were those involved aware of the plan? What was the balance between prepared response and improvisation? And most important, how did the Department of Energy respond to the Three Mile Island accident in the spring of 1979?

The historian writing the history of the Department of Energy at Three Mile Island finds himself in the position of telling a largely unknown story. Given the significance of the Department's activities, this requires some explanation. The Department's airport command post was isolated from reporters, and many government officials were unaware that the Department of Energy was even present. This isolation was deliberate. The Department of Energy consistently muted its presence throughout the hectic days of the crisis and avoided making any public statement about its activities. As a result, what it did remained unknown to most Americans.

The historian thus finds himself called upon to use traditional skills: telling the story, distinguishing the significant from the insignificant, comparing planned or intended actions with what actually happened, and making known the unknown. Our task was to create order out of chaos—from the various log books, notes, oral interviews, and radiological data we assembled—and in doing so, we created a Department of Energy archive on Three Mile Island.

A major task we faced was selection. The Department of Energy archives, combined with the records of the Nuclear Regulatory Commission, the Kemeny Commission, and other agencies are vast. Any historian of the twentieth century is familiar with the problem of dealing with too much data, not too little. In selecting which data to use and what stories to tell we have tried to be reasonably comprehensive, but also to stress stories of significant human interest. For the story of the Three Mile Island is much more than a story of nuclear technology. It is a story of human beings making history.

The human element was especially crucial at Three Mile Island. People often responded not to what was actually happening but to what they feared might happen, and their apprehension, unfounded as it turned out to be, affected decisions in the present. There was no sharp division between technical and human problems. Mechanical failures were magnified, even created by human decisions or indecision, and technical data often got lost in bureaucratic entanglements. Scientists did not know where their technical data were going; meanwhile politicians were making decisions without considering technical data. Communications between expert and public, between scientist and politician, and between the center and the periphery were a constant problem. At almost every point in the action, the key problems were not technical but human.

To emphasize the human element in the story of Three Mile Island we have conducted a number of oral interviews with those who responded to the nuclear accident. Their recollections have provided a valuable supplement to the written evidence, for these people lived through a time of confusion, exhaustion, and fear that is not often reflected in the terse prose of bureaucratic documents. If Three Mile Island resembles a military battle, these were its veterans. Their testimony provided vital new evidence and different perspectives, and they helped reinforce our conviction that Three Mile Island is an important story of human and institutional behavior, as well as of nuclear technology.

☆ ☆ ☆ ☆ ☆ ☆ ☆ ☆ ☆ ☆ ☆ ☆ ☆ ☆

This history is the result of cooperation between two historians and a government agency. On one level we have told the story of that agency's role in responding to a nuclear accident. We have also tried to place the Department of Energy in the broader context of the total response to the accident. In addition, we have kept in mind the fact that Three Mile Island was a battle in both a metaphorical and a literal sense. The Department of Energy teams, which had prior military experience with weapons-related accidents, often resembled an army engaged in the struggle against an invisible enemy, radioactivity. And as generals constantly learn from past battles, so the lessons of Three Mile Island are still being learned.

We are grateful to the Department of Energy for making accessible its archival holdings on Three Mile Island, and to the many employees and contractors who in all interviews helped the documents come alive and facilitated our understanding of a complex historical event. In particular we would like to acknowledge the assistance of Ruth C. Clusen, Assistant Secretary for Environment; L. Joe Deal, Herbert F. Hahn, Robert Friess, Jack M. Holl,

Introduction

Richard G. Hewlett, Rodney P. Carlisle, and Roger M. Anders of the Department of Energy; George Mazuzan and Stephen Scott of the Nuclear Regulatory Commission; Lee Johnson of the National Archives and Records Service; G. Robert Shipman, John F. Doyle, and Robert Meibaum of EG&G, Inc.; Bruce K. Segal, Ann Gilbert, and Romuald J. Misiunas of C&W Associates/Historical Consultants worked tirelessly at various stages of our work and kept it going through difficult times. Patricia Garrison typed much of the manuscript, with skill and good humor. To Edith Notman, our editor and critic, and most important, an understanding and patient colleague, we are most grateful. We would also like to thank Ann K. Williams and Washington University in St. Louis for understanding and support in many ways. None of them should be held responsible for any sins of omission or commission on our part.

St. Louis   P.L.C.
June 1981   R.C.W.

Chapter I
Accident at Three Mile Island

*The basic philosophy was: "Everything is go-
ing to work and it will be O.K."*

Ed Frederick

*I think we knew we were experiencing some-
thing different, but I think each time we made
a decision, it was based on something we
knew about.*

Gary Miller

The familiar sight and sound alarmed no one. The time was four a.m.,
the date March 28, 1979. Technology seemed under control.
Michael Donelan, handing out spare parts at a warehouse in the Unit
2 complex, had heard and seen it before and thought nothing of it. A mile
across the Susquehanna River in Goldsboro, John and Holly Garnish
awakened that night to a familiar sound, and went back to sleep. Bill Whittock,
a retired civil engineer, stumbled out of bed and peered across the river at a
rising column of white steam. Most residents of the Harrisburg, Pennsylvania,
area, including those who had seen the film *The China Syndrome* the night
before, were asleep, unaware that anything was happening at the plant. Yet in
a few days the cooling towers at Three Mile Island would come to symbolize
the dangers of America's most serious nuclear power plant accident, an accident that would alter the perceptions and the directions of nuclear energy in
the United States. But to nearby residents, the Unit 2 nuclear power plant on
Three Mile Island was simply venting steam to the atmosphere.[1]

To the men inside the Unit 2 control room on March 28 the initial events of
what came to be known as the Three Mile Island nuclear accident were as or-
dinary as the steam venting observed by Mike Donelan, Bill Whittock and the
Garnishes. Unit 2 plant operators were accustomed to constant alarms and
signals during the course of reactor operation, and many had occurred in the
year since March 28, 1978, when Unit 2 had first begun to operate at full
power. The operators the morning of the accident were the graveyard shift,

coming on at eleven in the evening and going home at seven the next morning. All were former Navy men from the nuclear program of Admiral Hyman G. Rickover. Operators Craig Faust and Ed Frederick had each spent about six years on nuclear submarines before coming to work for Metropolitan Edison, the utility based in Reading, Pennsylvania, that operated the Three Mile Island plant. The shift supervisor, Bill Zewe, also had six years of experience with naval reactors, and the shift foreman, Fred Scheimann, had worked for eight years as an electrical engineer on similar submarines.[2]

On the night of March 27-28 these four men and six other technicians were operating the Unit 2 nuclear reactor at ninety-seven percent of full power. The night marked the first anniversary of the reactor's operation, which might have been cause for a small celebration. It was not.[3]

Not everything was running smoothly. For about eleven hours successive crews had been trying to unclog the Unit 2 demineralizer system. The demineralizer system removes silt and dissolved salts from river water before the water can enter the reactor piping; if the water does not flow freely, the system automatically "trips," or shuts off. To clear out some resin that had collected in the demineralizer system, workers had been using high-pressure streams of air and water. At about four a.m., the system became completely clogged, tripping the main feedwater pumps, and finally the huge turbine of Unit 2.[4]

The first response of the system was completely automatic. The malfunction triggered a series of mechanical actions to shut down the reactor. The turbine trip immediately produced a buildup of steam pressure in the reactor coolant system, automatically opening an electromatic relief valve and allowing steam to escape. Simultaneously the reactor shut down, or "scrammed," within eight seconds as control rods dropped into the reactor core to absorb the neutrons flowing between columns of uranium fuel pellets sheathed in silver-colored tubes of zirconium.

Flashing lights and ringing alarms filled the Control Room. "As soon as I seen that," recalled Zewe, "I jumped up and just as I got through the door, I looked and I seen that we had a turbine trip and I took another step or so and got up to about the first few desks there and then the reactor tripped."[5]

"Oh my God," Ed Frederick reportedly exclaimed, "this is a classic."[6]

At this point the first mechanical problem developed. Although the reactor coolant pressure now dropped as anticipated, the electromatic pressurizer relief valve did not reseat as it should have. Instead it remained open, allowing reactor coolant water to escape into the containment building and then into the auxiliary building. The operators could not have known that for about two hours coolant water circulated through the reactor only to spill into the plant.

A combination of mechanical and human errors compounded the initial problem. When the reactor had scrammed, an emergency core cooling system had automatically begun to inject high-pressure streams of water into the reactor core. Control room instruments indicated that this system was working properly, when in fact it was not. The instruments showed that water was rising in the reactor core. However, the operators had no way of knowing that two valves in the emergency feedwater lines running to the reactor core had been closed two days earlier for routine maintenance — and were still closed.[7] No water reached the core by this route, but the instruments did not reflect that fact.

Lights on the Unit 2 control room instrument panel indicated that these valves were shut, but for eight minutes their warnings went unheeded. A large yellow caution tag hanging on the panel obscured one light. An operator blocked the other.[8] Instruments indicated that too much water was flooding the reactor core, when in fact the closed valves blocked the route.

Next a human decision further exacerbated mechanical problems. Two minutes after the start of the accident, the emergency core cooling system started up. A minute later, Craig Faust decided that he was injecting too much water into the core, pushed six buttons that gave him manual control of the system, and throttled back the flow of water. He did this in part because a pressure gauge had risen off scale, implying that both the steam generator and the pressurizer, a device that enabled operators to control pressure in the reactor, were full of water. In fact, because of the two closed feedwater valves, the steam generator was not flooded, but was running dry. At 4:04 and 4:10 a.m. Faust manually throttled back two pumps used to keep the reactor at a safe coolant level, attempting to solve a problem that did not exist. Here, as throughout the accident, the operators feared "going solid" — having too much water in the system and losing control of pressure — when they should have feared too little water and an uncovered reactor core.

Eight minutes into the accident (Faust later said it seemed like only sixty seconds) the problems began to multiply. When operators identified the two closed feedwater valves and opened them, they were promptly confronted with a Christmas-tree effect on the Unit 2 control room instrument panel as hundreds of red and green lights flashed and alarms sounded. "I would have liked to have thrown away the alarm panel," Faust later recalled; "it wasn't giving us any useful information." Having opened the feedwater valves, Faust and Scheimann observed that within another three minutes the level on the pressurizer gauge had dropped back on scale and was actually decreasing. Next they restarted the pumps in the emergency core coolant system — and created still another crisis for themselves. Now water flowed through the two newly opened valves, continued through the stuck electromatic relief valve and out into a tank designed to hold any overflow. However, deluged by thousands of gallons of radioactive water and steam, the rupture disk in the tank simply burst under the increased pressure. More than eight thousand gallons of water began rushing out of the containment building and onto the floor of the auxiliary building.[9]

In a report issued several months after the accident, the Nuclear Regulatory Commission stated that, given these conditions, Unit 2 operators should have declared a Site Emergency at the Three Mile Island plant at 4:15 a.m. According to the Metropolitan Edison Emergency Plan, a Site Emergency exists when there is a loss of primary coolant pressure, or increased pressure in the reactor building or high levels of water in the reactor building sump. At 4:11 a.m. there had indeed been a high-water-level alarm. By 4:15 reactor coolant pressure had dropped below the point where the emergency core cooling system was activated, and reactor building pressure had risen. Yet no Site Emergency was declared, because the operators believed that the drop in reactor pressure was under control, reactor building pressure seemed slight, and there was no evidence of any radiation release.[10]

The reactor refused to behave. At 4:38 operators finally realized that water was now escaping from the containment building and turned off the sump pumps. But it was too late. The auxiliary building sump tank had overflowed into the auxiliary building sump, and radioactive water had backed up through floor drains onto the floors of the auxiliary and fuel-handling buildings.[11] By five a.m. technicians were monitoring increasing background radiation levels. At 5:18 the first of a series of radiation alarms sounded, indicating excessive radiation levels in the air of the containment building. To complicate matters further, paper jammed in an alarm printer in the Unit 2 control room and a fire alarm sounded.

"At that point," recalled Ed Frederick, "I felt it was increasingly difficult to determine the actual conditions of the reactor cooling system from the indications that were present on the panel."[12] Meanwhile, Faust and Zewe continued to shut down all coolant system pumps in an effort to restore the air bubble in the pressurizer.

This was the situation at about 5:45 that morning when Joe Logan, the Unit 2 Station Superintendent and a twenty-five-year veteran of nuclear submarines, reached the plant. "What I saw when I arrived," said Logan, "was extremely difficult to fathom; . . . we had no pumps running, we had a steam generator that was isolated, we had a high pressurizer level which to me at that time indicated that we had the core covered; we had a low pressure, which didn't go along with the other indications."[13]

Radiation levels inside the plant were becoming a major concern. At 5:40 Dick Dubiel, Radiation Protection and Chemistry Supervisor, arrived and immediately ordered an air sample taken from the containment building. "We were trying to determine the activity levels in containment atmosphere," Dubiel later recalled; "we had difficulty because the atmospheric monitor was in fact flooded with water which in fact was I believe steam condensing in the sample lines."[14] The sample contained traces of cobalt-60 and the noble gases xenon and krypton, suggesting a leak from the primary into the secondary coolant system. Dubiel also directed that a sample be taken directly from the reactor coolant lines. A few minutes later a radiation/chemistry technician was trying to change a filter in the containment air sample monitor. It blew out under pressure, indicating a buildup of steam inside the containment building.

Increasing radiation implied possible fuel damage and leakage from containment. Shortly after six that morning plant managers discussed the situation in a 38-minute conference telephone call arranged among the homes of several officials and the plant. On the line were Gary Miller, the Unit 2 Station Manager; George Kunder, the Unit 2 Superintendent for Technical Support; John Herbein, the Metropolitan Edison Vice President for Generation; and Lee Rogers, the site representative for Babcock and Wilcox, the builders of Unit 2. Miller summarized the call:

> Following some discussion of the conditions it was agreed we must believe our instruments. It was also noted by the plant that the reac-

tor coolant pumps were presently off due to a loss of flow indication, the electromatic valve was thought to be shut, and there was no indication of radiation at this time. Additionally, the rupture disc on the reactor coolant drain tank in the Reactor Building was noted to have been ruptured and, therefore, we were aware that there was some water within the Reactor Building sump.[15]

In fact, there *were* already indications of radiation. Moreover, the ruptured disk had flooded building floors, which surprised no one because it had happened before. The men discussed the electromatic relief valve during their conference call but did not discuss the two feedwater valves that had been shut during the first eight minutes of the accident.[16]

At the time of the conference call, the reactor core was partly uncovered. Temperature readings inside the core later turned out to be extremely high. Only after eight a.m. did plant operators realize that these temperatures had reached 2620° F, raising the possibility of some damage to the zircalloy fuel cladding, the release of fission products, and the generation of hydrogen.[17] None of this was known with any certainty at the time.

One consequence of the partially uncovered reactor core was the production of increasingly high radiation levels inside the plant. At 6:24 a.m., the reactor building radiation monitor sounded an alarm. Two minutes later the area monitor in the reactor building also went off, indicating excessive radiation levels. Most important, at 6:35 the containment dome radiation monitor alerted the control room to additional radiation problems. For the second time that morning a condition had been reached which, according to the Nuclear Regulatory Commission, should have resulted in the declaration of a Site Emergency. Again, it did not.[18]

Instead, Unit 2 personnel began surveying radiation levels in the plant. Between six-thirty and seven a.m. Mike Janowski, a radiation/chemistry technician, walked through the Unit 2 auxiliary building with a portable beta-gamma survey meter and reported rapidly increasing radiation levels. Dale Laudermilch, an auxiliary operator, later recalled that he was in the auxiliary building "trying to figure out where the devil is this water coming from" when Janowski ran down the hall yelling, "Get the hell out!" At 6:48 more monitor alarms sounded, this time from the vent stack particulate monitor and the machine shop in Unit 1, through which ran some unshielded reactor coolant sample lines. "This was very alarming to me,"recalled Dubiel; "I had thought we had some leaking fuel rods and it was apparently substantial."[19]

Nearly three hours after the start of the accident, Unit 2 declared a Site Emergency. At 6:48 a.m. an announcement over the plant public address system ordered all personnel out of the auxiliary building. At 6:55 Bill Zewe declared a Site Emergency, based on the alarms sounded by radiation monitors throughout the plant. At 7:02 Zewe placed a telephone call to the Pennsylvania Emergency Management Agency, the state civil defense organization. Zewe informed the duty officer of the Site Emergency and requested that he notify the Pennsylvania Bureau of Radiation Protection.

For the first time since the start of the accident at Three Mile Island someone outside the plant knew that the events of the previous three hours were not merely something unusual, but an emergency.

Bill Whittock later claimed he, too, had noticed something unusual. Stepping out his front door, he had noticed a strange metallic taste in the cool air of the early spring morning. He had wanted to call Civil Defense but could not find any listing in the telephone directory between Civil War Times Illustrated and Civil Air Patrol Squadron. Down at the marina across the river from the plant a young boy excitedly told him that a helicopter had landed in a nearby field. Whittock walked over to take a look and discovered a television news team, which promptly interviewed him.[20]

Bill Whittock had roused himself at four a.m. to witness the perfectly normal sight of Unit 2 venting steam over Three Mile Island, because as an engineer he found it interesting. He had gone back to sleep. Later that morning he found himself on camera as the first witness to what he was told were dramatic and alarming events. The scope of the accident was beginning to escalate both on and off Three Mile Island.

Chapter II
Plans and Portents

In spite of what may appear to be adequate control measures, a reactor failure is always a possibility. A complete well-conceived disaster plan, including evacuation, treatment of injured and decontamination must be set up. Failure to do this can lead to chaos, panic and increased damage in the event of an accident.

Charles R. Williams
June 1956

Gently rolling farmland cradles the Susquehanna River below Harrisburg, Pennsylvania. Southeast of the state capital, about ten miles downstream, the river veers sharply, turning nearly due south as it flows past the small borough of Middletown. Here the Susquehanna widens and turns sluggish, as if reluctant to push itself the remaining miles to the Chesapeake Bay. Seemingly exhausted, it sheds part of its burden of silt and in doing so has created over the centuries the many islands which punctuate its course to the sea. Events have focused world attention on the largest of these islands.[1]

Three Mile Island. The origin of the name is obvious, Probably too obvious. Snuggled against the eastern shore of the river, the island is almost exactly three miles long. As ice age glaciers receded thousands of years ago, they deposited huge boulders in the wide channel of the river. Around these boulders river sediment collected, slowly building up the 625 acres which would become Three Mile Island.[2]

Evidence of past civilizations using Three Mile Island abounds. Archeological digs indicate that the island was a popular hunting and fishing site nearly four thousand years ago. Recent excavations have also uncovered shards of flat-bottomed soapstone bowls and clay pottery from as early as 1700 B.C. Stone chipped knives, spear points, arrowheads and other stone tools produced by the Susquehannock Indians over a thousand years ago have also been unearthed.[3]

Present civilization has left its mark as well. Before construction began on the first unit of the power plant in 1968, Metropolitan Edison used the island for agriculture and recreation. The company leased over half the island to local farmers who planted tomatoes and corn, and the cornfields provided food and shelter for a sizeable flock of ring-necked pheasants. The company also maintained a small picnic area with tables, fireplaces and a well. At another spot, the utility built a boat dock for sailors and fishermen. At the southern end of the island a wooded area of black locust trees provided a haven for white-tailed deer and other wildlife. Today, the trees still stand, but the fields and the picnic tables and the dock are gone. Instead, giant cooling towers, symbols of modern man's victory over the mysteries of nature, dominate the northern part of the island.[4]

The four 370-foot cooling towers of Metropolitan Edison's Three Mile Island nuclear power station loom above the land much as a medieval castle towered over its surrounding countryside six hundred years ago. The analogy, if not perfect, is apt. Both structures project an aura of awesome power. To the serf of the fourteenth century, power and protection flowed from the military strength and legal authority held by the lord of the castle. To twentieth-century man, the power station produces vast amounts of electrical power to light his home, run his appliances and speed his communications: in short, to protect his standard of living. In the past the failure of a castle threatened the lives of those it was designed to protect. Today, the failure of a nuclear power plant presents a similar danger. Should the power locked within the walls of either structure become uncontrollable and escape whether in the form of armed men bent on pillage and bloodshed or an invisible cloud of radioactivity, the surrounding countryside would take years to recover.

Fourteenth-century man and his twentieth-century descendant share a common fear: the unknown. In the fourteenth century the great fear was the terror of the Black Death, bubonic plague. Striking without warning, making no class distinctions, the plague drove frightened people from their homes into the countryside or the mountains to escape its agonizing grip. None understood the plague's origin, none could see its cause. In this sense, radioactivity is twentieth-century man's Black Death. It is invisible. It is terrifying. It is imperfectly understood. It does not differentiate among its victims. It is believed by many laymen to be contagious. Like the plague, some survive exposure to radiation while others succumb. But even more terrifying than fourteenth-century man's blind battle with plague, excessive radiation can strike down even those who survive, by affecting their ability to produce genetically sound children.

The economic relationships between serf and castle, citizen and company, are different, however. Whereas the medieval serf paid part of his annual harvest to the lord of the castle for the use of the land and protection, those who lived in the shadow of the proposed Three Mile Island station viewed the power plant as provider to the community. The towns surrounding the island — Middletown, Highspire, Goldsboro, York Haven, Falmouth and Cly — had already benefited from the millions of dollars of tax revenues and the increased employment opportunities that the new plant generated. The closing of Olmstead Air Force Base in 1967 had significantly clouded the economic life of the region and the Three Mile Island nuclear power plant was widely regarded as a remedy. Perhaps the cloud *did* have a silver lining. [5]

By 1966 Metropolitan Edison had reached a decision to build a nuclear reactor at Three Mile Island. Two years later, in December of 1968, the company decided to construct a second reactor, Unit 2, on the island, arguing that electricity could be produced more quickly and more efficiently by building the unit at Three Mile Island rather than another site. [6]

For Metropolitan Edison and its parent company, General Public Utilities, the main reason for building the plant on Three Mile Island was the cost. The island site, according to the utility's report to the Atomic Energy Commission, was "economically advantageous." Power transmission lines already existed and could be used while the older, less efficient coal-fired plants in the vicinity were being phased out. The Pennsylvania Turnpike and the Penn Central Railroad were both convenient. Another factor, the report continued, was the steady corporate market for electricity represented by the large Bethlehem Steel plant at Steelton and the giant Hershey Chocolate Company complex in Hershey. In addition, cooling water for the reactor was readily available from the Susquehanna River at a lower cost than at any of the alternative sites considered by the utility. A Metropolitan Edison feasibility report further pointed out that construction wage rates were low and worker productivity high in the area surrounding Three Mile Island. And finally, because the company had owned the island since the early nineteen hundreds, no acquisition costs were involved. Persuaded, the Atomic Energy Commission agreed to license Three Mile Island Unit 2. [7]

At the center of Three Mile Island Unit 2 stands the containment building housing the pressurized-water reactor designed and built by Babcock and Wilcox. Unit 2 was licensed on February 2, 1978, and began commercial operation on December 30. Its sister, Unit 1, was licensed for commerical power operation on April 19, 1974, and began commercial operation on September 2. When the accident occurred, Unit 1 was undergoing routine maintenance. In early 1979 both plants provided a much needed source of energy for southern Pennsylvania.[8]

Inside the containment building of the Unit 2 reactor is the cylindrical nuclear core, a steel vessel in which hang 177 tubular fuel assemblies. Each fuel assembly contains 208 enriched uranium fuel pellets, a total of 36,816. Between the fuel assemblies stand control rods which can absorb neutrons and determine the rate of fission. If the control rods are pulled up out of the assembly, more neutrons become available to split the uranium atoms. The more atoms that are split, the more heat is generated. If the control rods are let down into the fuel assemblies, they absorb neutrons, reduce fission, and lower the temperature in the reactor core. In an emergency the control rods are automatically lowered so that the reactor "scrams," or shuts down.

Because of a nuclear power reactor's complexity, the possibilities for malfunction are numerous. Many involve the piping, valves, seals, wiring and switches common to any power plant. In the licensing and early operating phase for any nuclear plant, all malfunctions, or "events," must be reported to the Nuclear Regulatory Commission. The Final Safety Analysis Report for Unit 2 at Three Mile Island anticipated a number of possible events, ranging from valve failures to earthquakes. The report did not anticipate the problems that a clogged demineralizer might cause, noting only that such a failure would not be "vital to the safe shutdown of the plant."[9]

Many events similar to those causing the accident had in fact occurred at Unit 2 before March 28, 1979. On October 19, 1977, water entered the air lines of the demineralizer, and the air lines had to be cleared out. When this happened again in May 1978, air driers were installed. On March 29, 1978, the electromatic relief valve failed to close for four minutes when a fuse blew. The reactor tripped and the emergency core cooling system functioned properly, but operators could not understand what had caused depressurization in the reactor coolant system because there was no indicator in the control room to show whether the valve was open or closed. Unit 2 personnel later admitted the "there had been a concern that the block valve could stick shut or open if used too often."[10]

More problems followed. On November 3, 1978, Unit 2 was operating at ninety percent power when a technician mistakenly caused a loss of feedwater. When the main feedwater pumps tripped, pressure built up in the steam lines but was properly released by the electromatic relief valve. As the plant continued gearing up for commercial operation in late 1978, still other pump and valve failures occurred. In each case, backup systems worked and no radiation escaped. Consequently Unit 2 received its operating license and began to generate power on December 30, 1978.[11]

In early 1979 still more problems arose with Unit 2. On February 8 the diaphragm on the waste tank rupture disk broke and as of the summer of 1979 it had not been replaced. In the control room the paper in the alarm typer, which prints out a record of reactor problems, had a history of jamming, tearing and running too slowly to affect operations. On March 26, 1979, a routine test on valves in the emergency core cooling system created a major problem that would complicate the accident. The valves were closed during the test. Maintenance men later testified that they then reopened them after the test. But forty-two hours later, in the early morning of March 28, the valves were still closed, preventing water from reaching the core. The men also noted that they had at other times found valves mispositioned; in one case the same valves were found closed "for some kind of test" but "the valves weren't reopened when they were supposed to be." Finally, an operator said that on one or two separate occasions a turbine had tripped because of "problems with the condensate polishing system."[12]

The problems that occurred at Three Mile Island were neither unique to that plant nor unanticipated at the time of the accident. According to Harold Denton of the Nuclear Regulatory Commission, pressurizer relief valves were a recurring problem at other plants; they had stuck open about a hundred and fifty times on Babcock and Wilcox reactors. In 1977 at the Davis-Besse plant in Toledo, Ohio, a thousand gallons of water spilled into the containment building when the same type of valve stuck open. More than ten months before the Three Mile Island accident, a Tennessee Valley Authority safety official, Carl Michelson, had called the problem to Babcock and Wilcox's attention in a report. Babcock and Wilcox engineers Bert M. Dunn and Joseph J. Kelly, Jr., had also written memoranda in the winter of 1977-1978 urging action on the pressurizer relief valve, warning that premature shutoff of emergency pumps and valves could also affect nuclear power plant accidents by misleading the plant operators during the accident. Finally, T. M. Novak, the chief of the Reactor Systems Branch of the Nuclear Regulatory Commission, had also warned in January 1978 that if the pressurizer relief valve in a Babcock and Wilcox reactor should stick open, then a loss of primary coolant water "might not be indicated by pressurizer level" as seen by control room operators. Operators might then erroneously shut off emergency coolant water just when it was needed.[13]

Medieval castles were designed to offer protection to a feudal society in time of war, plague or famine. They were man-made citadels defending their inhabitants from ravages of nature and society. But at Three Mile Island the danger lay within, part of the risk of extracting nature's power through man's technology.

☆ ☆ ☆ ☆ ☆ ☆ ☆ ☆ ☆ ☆ ☆ ☆ ☆ ☆

Like the plagues which periodically ravaged medieval Europe, nuclear radiation presents an invisible and deadly threat little understood by society at large. But it is a threat posed by the Promethean aspirations of man himself, not by nature. It can be deadly but it is not contagious, whatever the public may believe. Nuclear power experts must anticipate the dangers radiation can present, for the industry is required by law to prepare for even the most improbable nuclear emergency.

As part of its Final Safety Analysis Report, Metropolitan Edison was required to submit a "Site Emergency Plan." According to the plan, in case of emergency a utility Emergency Planning Group would immediately take over control of the reactor and serve as a liaison with state and local agencies. Medical facilities would be set up at the nearby Hershey Medical Center. The utility's own Radiation Protection staff would be available for help, as would resources of the Radiation Management Corporation of Philadelphia.

Like Dante's circles of Hell, the emergency contingencies of Metropolitan Edison rippled outward from center to periphery. A personnel or local emergency involved potential or actual hazards to individuals within the inner circle of the plant itself. A site emergency would be declared in case of "an incident which would potentially result in an uncontrolled release of radioactivity to the immediate environment" and possible off-site radiological dosage to the local population. A general emergency was defined as "an incident which has the potential for serious radiological consequences to the health and safety of the general public."[14]

According to the plan, a site emergency could be triggered by high radiation levels, radiation alarms or the loss of primary coolant pressure. If one were declared, plant personnel would evacuate affected buildings and notify the State of Pennsylvania and the Nuclear Regulatory Commission. What actually led to the declaration of a site emergency on March 28, 1979, were high radiation levels in the reactor building.

A general emergency required the same actions as a site emergency, plus immediate off-site monitoring beyond the island itself. The Final Safety Analysis Report for Unit 2 noted that serious accidents might occur, even though they were extremely unlikely. Serious accidents could include various types of loss-of-coolant accidents that might cause fuel rod damage, fuel melting or radiation leakage from the containment building to the environment. The condition in the plant which led to the general emergency on March 28 was the sounding of the reactor building high-range gamma monitor *high* alarm, indicating excessive radiation in containment.

In any emergency Metropolitan Edison personnel were to be assigned specific tasks. The Unit 2 Station Superintendent would immediately become an Emergency Director implementing a Radiation Emergency Plan. Drills of radiological monitoring teams, fire brigades, repair parties, and first aid teams had been held before the accident. But none of these drills had anticipated the exact nature or severity of the events of March 28.[15]

Prior planning for a nuclear emergency had also included agencies of the Commonwealth of Pennsylvania. According to the Unit 2 Final Safety Analysis Report, in case of emergency Metropolitan Edison was responsible for the "prompt notification" of Pennsylvania authorities, the Bureau of Radiological Health and Pennsylvania Emergency Management Agency. The Bureau of Radiological Health would then be responsible for "the management of all off-site aspects of a radiation emergency" and the Pennsylvania Emergency Management Agency would carry out "the required protective actions." By March 1979 the Bureau of Radiological Health had been renamed. Called the Bureau of Radiation Protection, it was directed by Thomas M. Gerusky. Reporting to Gerusky was the chief of the Division of Nuclear Reactor Review and Environmental Surveillance, Margaret A. Reilly, and her staff, including nuclear engineer William P. Dornsife. The director of the Pennsylvania Emergency Management Agency was Colonel Oran K. Henderson, who had formerly commanded American troops in Vietnam.[16]

Pennsylvania also had an emergency plan for nuclear power generating stations. The Bureau of Radiation Protection plan described many contingencies involving nuclear plant accidents and off-site radiation exposure, dividing them into three different classes: industrial accidents involving plant personnel but "little or no off site radiological impact;" incidents involving a threat to public health and safety, security violations and minor radiation releases; and incidents creating a direct radiation exposure hazard to off-site populations.[17]

For all incidents, Pennsylvania required that the plant notify both the Bureau of Radiation Protection and the Pennsylvania Emergency Management Agency, but not the surrounding towns and counties. Once the Pennsylvania Emergency Management Agency learned of an accident, it would inform county civil defense agencies, which in turn would "be responsible for advising county and local government officials of those events which are indicated as being of potential public interest."[18] In the event of a serious accident, the Pennsylvania Emergency Management Agency would activate a State Emergency Operations Center, and with Bureau of Radiation Protection guidance, broadcast emergency information by radio and television. It would also notify the Governor and Lieutenant Governor of Pennsylvania of the accident and of its subsequent activities.

Under the plan the Bureau would alert state and local officials to radiation dangers and the need for protective action through the communications facilities of the Pennsylvania Emergency Management Agency. The Bureau was also responsible for "the notification of and the requesting of assistance from those federal agencies having intrinsic interest and expertise in radiation protection," including the Department of Energy.[19] The Bureau had the option of requesting any necessary assistance from the Department of Energy, including "aerial surveillance for deposited radioactivity."[20]

At the state level, then, in case of emergency the Bureau of Radiation Protection would conduct radiation monitoring, and the Pennsylvania Emergency Management Agency would direct communications and, if needed, evacuation of the local population. There was also a Commonwealth of Pennsylvania plan for an emergency at Three Mile Island itself. Anticipated accidents included the unplanned release of radioactive waste into the Susquehanna River, potential release to the atmosphere and actual release to the atmosphere. Should radiation escape, the Bureau of Radiation Protection would perform off-site monitoring for radiation and would advise both the State Police and county civil defense officials if emergency measures, such as "take cover" advisories or alerts, were to become necessary.[21]

In case of a major accident, Metropolitan Edison was to contact the Pennsylvania Emergency Management Agency which would in turn contact the Bureau of Radiation Protection, which would then telephone the control room at Three Mile Island to determine the extent of any radiological danger. The Bureau would subsequently advise the Governor if he should declare a state of emergency or evacuate the area.

The federal government, notably the Nuclear Regulatory Commission and the Department of Energy, stood at the outer circle of emergency preparedness. The Final Safety Analysis Report for Unit 2 noted that because the Department specialized in radiation safety and medicine, it could provide "valuable assistance in case of an emergency." The Report also observed that while the Nuclear Regulatory Commission was primarily a regulatory and investigatory agency, it might "render emergency assistance."[22] The State relied on federal agencies for help with radiological problems, and the Department of Energy was the primary agency expected to give assistance.

☆ ☆ ☆ ☆ ☆ ☆ ☆ ☆ ☆ ☆ ☆ ☆ ☆ ☆ ☆ ☆ ☆

In the summer of 1976, nearly two years after Three Mile Island's Unit 1 began producing commercial electricity, a small silver and blue airplane, loaded with sophisticated electronic gear, eased off the runway at the Capital City Airport near Harrisburg, rose in easy circles toward the east and headed down the Susquehanna River. Inside the twin-engine Beechcraft the electronic hardware began to measure and map the levels of background radiation in the area surrounding the power station. The flight was part of a program, begun by the Atomic Energy Commission and continued under the Nuclear Regulatory Commission, to survey natural radiation levels around facilities handling nuclear materials. The crew and equipment were part of a large Department of Energy radiation assistance program.[23]

The scientists and technicians conducting the survey were in a sense the products of nearly twenty-five years of governmental concern with radiation safety. The concern had grown out of the weapons testing program in the early 1950s, especially after the extensive fallout from the 1954 Bravo test in the Marshall Islands. After the passage of the Atomic Energy Act of 1954, which encouraged the commercial development of nuclear energy, sites handling nuclear materials were no longer under direct control of the Atomic Energy Commission. The fact that the materials were more widely distributed meant that the probability of an accident which would threaten the public health and safety rose significantly.

Initially developed to measure weapons fallout, aerial monitoring of radioactivity was soon applied to civilian projects, including nuclear power stations. The application of military techniques to civilian situations was practical and economic. Ideally, surveys could be flown to determine levels of natural radioactivity in the area surrounding a proposed nuclear site. Then, after the plant had begun operation, the area could be resurveyed to detect any changes. Moreover, the funds spent to keep the monitoring teams in readiness would not be wasted between military tests, for the scientists could continue to hone their skills and their instruments on civilian projects. The direct ancestor of the silver-and-blue Beechcraft flying over Three Mile Island was a lumbering DC-3 equipped with radiation detection instruments. Operated by the United States Geological Survey, the plane had been used to locate uranium deposits from the air and also to measure fallout from the Nevada weapons tests in the 1950s. In 1958 the Atomic Energy Commission's Division of Biology and Medicine, recognizing the value of such aerial measurements following the spread of contamination from a reactor accident at Windscale, England, acquired the DC-3 and began a nationwide program to measure radiation levels around nuclear facilities. But the cost of operating the DC-3 was too high, about $300 per flight hour, and the Commission decided to have the work done by an outside contractor.[24]

The next step was finding a contractor to do the work. In 1959, the Atomic Energy Commission approached Herbert Greer, a former professor at the Massachusetts Institute of Technology and the founder of EG&G, Inc., the company which had managed the Commission's Nevada operations. EG&G had had long experience at the weapons test site and could apply this to nonmilitary radiation monitoring. Consequently, in the early 1960s EG&G began surveying government nuclear facilities near Las Vegas, Nevada, Norfolk,

Virginia, and Aiken, South Carolina. The survey used sodium iodide crystals to determine radiation levels and a Doppler radar navigation system to pinpoint the aircraft's position. By combining the radiation results with the location of the plane when the measurement was taken, scientists could construct an accurate topographical map or "footprint" of radiation levels in the area. In 1967, the Aerial Radiation Measuring System (ARMS), as EG&G dubbed the survey operation, began to map the areas around commercial reactors. Nine years later, the same company surveyed the Three Mile Island power station.[25]

Between August 2 and 4, 1976, the team of EG&G scientists and technicians headed by Al Fritzsche surveyed a 625-square-mile area surrounding the Three Mile Island plant. Only the area directly over the station was omitted from the survey. The plant operator did not want to risk an aircraft crashing into the reactor, Fritzsche recalled, even though the containment buildings were designed to withstand the impact of a commercial airliner. This safety feature had been required due to the proximity of the plant to the Harrisburg International Airport. The crew flew across the twenty-five-mile grid at half-mile intervals at a height of five hundred feet. In addition, circular patterns with a one-mile radius were flown around the station at 500, 1000 and 2000 feet. At the same time a ground team collected soil samples from two sites east and southeast of the plant. The scientists found that the radioisotopes which were detected and the gamma ray exposure rates associated with those isotopes were consistent with the expected normal background levels.[26]

Once the field survey was completed, the EG&G team headed back to Las Vegas to prepare maps illustrating the radiation levels around Three Mile Island and draft a final report of their findings. More than two years later, these maps would play a critical role in flights conducted by other EG&G aerial monitoring teams following the accident at Three Mile Island.

The aerial surveys were only part of the Atomic Energy Commission's radiation safety program. In 1958 and 1959 each of the contractor-operated national laboratories and testing sites had organized groups of qualified scientists and technicians into radiological assistance teams. Not only would these teams respond immediately to an accident at their particular national laboratory or site, but they would also travel to the scene of an accident if a source of radioactivity were endangering public health and safety. Any citizen could request radiological assistance and help with a nuclear-related problem, ranging from a lost radioactive needle used in a hospital to a malfunction at a nuclear reactor. Should a truck carrying radioactive isotopes be involved in an accident and the materials scattered, a radiological assistance team would move in, survey the area, locate the radioactivity, advise local officials and insure that no one was harmed.[27]

Radiological assistance teams were prepared to conduct a wide range of radiation monitoring. Using the latest field instruments and mobile vans equipped with computers, the teams could collect soil, air, water and vegetation samples and analyze them quickly in the field. Recognizing new needs, team members continually improved radiation measuring instruments. For example, in the months before the Three Mile Island accident, a scientist at the Brookhaven National Laboratory in Upton, Long Island, had developed a sophisticated filter which allowed an accurate measurement of the presence of radioactive iodine-131 in the air. Such measurements had been difficult previously because radioactive xenon and krypton, fission products which were present with iodine, emitted higher levels of gamma radiation, saturating existing instruments and allowing the iodine to go undetected. Since iodine-131 lodged in the thyroid gland unless preventive action were taken, this new instrumentation represented an important advance in protecting the public.[28]

Another recent development in radiological safety had taken place at the Lawrence Livermore Laboratory in Livermore, California. There, scientists combined computer technology with meteorology to predict possible levels and areas of radioactive fallout. The system was tagged with a convoluted name, Atmospheric Release Advisory Capability, but the acronym ARAC was much easier to handle and few knew the system by any other name. After locating the radioactive cloud or plume through aerial measuring, the ARAC could use the levels detected in the plume to calculate the radiation levels of the source. Then, plotting these measurements against weather conditions and the topography of the area, within five minutes the computer could tell ground monitoring teams where they should expect to find radioactivity. Not only did this system integrate aerial and land radiation monitoring efforts, but when used properly it could also provide invaluable aid in evacuation planning.[29]

Long before the crisis at Three Mile Island, then, these radiological assistance teams, under contract first to the Atomic Energy Commission, then the Energy Research and Development Administration, and finally the Department of Energy, were responding to radiation problems throughout the country. Most of the problems were quickly solved. A short phone conversation would pinpoint the difficulty and a team member would explain how the individual on the spot should handle the situation. At other times, a team would travel to the "hot spot" and work with authorities there to clear up the problem. A few responses took longer. A number of radiological assistance teams had spent several weeks in the Marshall Islands in the Pacific Ocean working with natives who had been caught in fallout from a weapons test in 1954. In spite of the diversity of their work, the radiological assistance teams viewed their task as a limited one: to determine if radiation levels were high enough to endanger public health and safety. If no problem existed, their job was over. If there were a danger, they would continue to take radiation measurements and furnish that information to local authorities.

The distinction between determining risks to public health and safety and making policy was an important one. Anyone could request radiological assistance from the Department of Energy, but neither the Department nor the contracting laboratories could respond unless asked. To scientists, technicians and Department of Energy officials, local authorities were always in command. The radiological assistance response was informational and advisory only. Policy decisions, while based on radiological assistance or aerial measurements, were made by others, such as private citizens, corporations, a state or another federal agency. In the end, these groups were responsible for actions taken to protect the public.[30]

Eleven federal agencies could respond to a nuclear crisis. Built around the Interagency Radiological Assistance Plan, the combined federal ability to respond to nuclear accident was long-standing, but untried.

Federal interagency cooperation on radiation monitoring, like the development of radiation assistance teams within the Atomic Energy Commission, grew out of the military experience. In November 1957, the Department of Defense and the Atomic Energy Commission began to develop formal emergency plans to cope with accidents involving atomic weapons. Four months later, in February 1958, the two agencies signed an agreement to coordinate radiation monitoring and medical safety efforts in responding to weapons accidents. In October the agreement was extended to include mutual assistance in any radiological emergency, military or civilian.[31]

Shortly after the Atomic Energy Commission and the Department of Defense agreed to combine their radiological knowledge, the Commission recognized that even this joint capability might not be adequate to cope with a major radiation accident. Consequently, a meeting with representatives of several federal agencies produced the Interagency Committee on Radiological Assistance, which operated on an *ad hoc* basis until 1961. In that year a formal Interagency Radiological Assistance Plan was signed by the Atomic Energy Commission, the Departments of Defense, Health, Education and Welfare, Labor, Treasury and Commerce, the Office of Civil Defense Mobilization, the Federal Aviation Administration, the National Aeronautics and Space Administration, the Post Office Department and the Interstate Commerce Commission. Since 1961, the Interagency Radiological Assistance Plan has been periodically revised, and the Environmental Protection Agency, the Nuclear Regulatory Commission and the Department of Agriculture have become participants.[32]

Under the Interagency Radiological Assistance Plan, successive nuclear energy agencies—the Atomic Energy Commission, the Energy Research and Development Administration and the Department of Energy—assumed primary responsibility for implementing and administering an emergency response in cooperation with other federal and state agencies. The Nuclear Regulatory Commission, created in 1975 when the Atomic Energy Commission's responsibilities were divided between the Nuclear Regulatory Commission and the Energy Research and Development Administration, was primarily responsible for licensing and inspecting reactors, and had neither the personnel nor the equipment to conduct radiation monitoring. Under a March 8, 1977, agreement between the Energy Research and Development Administration and the Nuclear Regulatory Commission, the latter would "notify the ERDA immediately of an emergency involving NRC licensees, facilities, or activities when NRC expects or anticipates that assistance and support from ERDA will be required." In return, the Energy Research and Development Administration was expected to "provide the resources of the Nuclear Emergency Search Team (NEST) and Aerial Radiological Measuring System (ARMS) assistance and radiological emergency assistance to support the Nuclear Regulatory Commission response to emergencies to the extent these

ERDA Radiological Assistance Regions &
Coordinating Offices

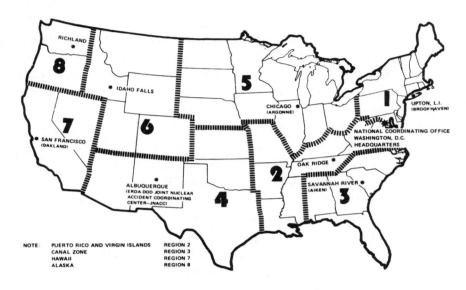

NOTE: PUERTO RICO AND VIRGIN ISLANDS   REGION 2
        CANAL ZONE   REGION 3
        HAWAII   REGION 7
        ALASKA   REGION 8

REGION	OFFICE	
1	Brookhaven	516 345-2200
2	Oak Ridge	615 483-8611
		Ext. 3-4510
3	Savannah River	803 824-6331
		Ext. 3333*
4	Albuquerque	505 264-4667
5	Chicago	312 739-7711
		Ext. 2111 Duty hours
		Ext. 4451 Off hours

REGION	OFFICE	
6	Idaho	208 526-0111
		Ext. 1515
7	San Francisco	415 273-4237
8	Richland	509 942-7381

capabilities are available and, with respect to radiological assistance, in accordance with the Interagency Radiological Assistance Plan."

The Nuclear Regulatory Commission-Energy Research and Development Administration agreement also provided that the Administration would undertake aerial radiological survey mapping of all nuclear sites, as had been done at Three Mile Island. When the Administration became part of the Department of Energy in October 1977, the agreement remained unchanged. When radiological monitoring, evacuation or medical assistance under the Interagency Radiological Assistance Plan was required, such help would be requested by local authorities through the regional field offices of the Nuclear Regulatory Commission or the Department of Energy. The Nuclear Regulatory Commission regional office at King of Prussia, near Philadelphia, and the Department of Energy office at the Brookhaven National Laboratory on Long Island had initial responsibility for Three Mile Island.[33]

In the event of a nuclear emergency, the Department of Energy's response would be governed in part by the "Emergency Planning, Preparedness, and Response Program," an Energy Research and Development Administration manual chapter written in the summer and fall of 1976. The program outlined the responsibilities of the Energy Research and Development Administration headquarters and regional staff as well as those of the Administration's contracting laboratories. When the Department of Energy assumed the Energy Research and Development Administration's responsibilities in 1977, the guidelines were not changed.[34]

Accordingly, when an emergency report came in to the Emergency Operations Center at the Department of Energy headquarters in Germantown, Maryland, the duty officer would obtain all the available facts and determine what additional actions should be taken. If the emergency appeared significant, such as an accidental release of radioactivity that presented a serious health hazard to the public, a group of senior Department of Energy officials known as the Emergency Action Coordinating Team would meet to discuss and coordinate the Department's response. Once a decision to respond was reached, the Emergency Action Coordinating Team would marshal the Department's radiation assistance resources.[35]

The manual clearly indicated that the original jurisdiction in a radiological incident at a commercial power reactor belonged to the individual or organization in control. However, it noted, other civil authorities such as local or state governments or another federal agency might take over. "Energy Research and Development Administration radiological assistance operations conducted on a site not under the jursidiction of the Energy Research and Development Administration are normally those emergency actions approved or requested by the individual organization in control . . . or an appropriate local civil authority.[36]

Once assistance had been requested for on-site accidents, the radiological assistance team closest to the scene would respond and, if necessary, establish a local command post to communicate with headquarters. The Department could quickly assemble equipment and alert personnel from the Aerial Measuring Systems operations in Las Vegas, Nevada and Andrews Air Force Base, Maryland. Additional radiological assistance teams from regional offices located in New York, Tennessee, South Carolina, Illinois, Washington, Idaho, New Mexico and California would also be alerted. Officials estimated that radiological assistance could be at the scene a few hours after notification.[37]

The equipment, the personnel, the guidelines for action were thus prepared and ready to respond to a nuclear accident before the spring of 1979. For twenty years radiological assistance teams and aerial monitoring crews had been honing their skills. It was upon this foundation that the Atomic Energy Commission, the Energy Research and Development Administration and the Department of Energy had constructed a continually evolving preparedness plan for a radiological emergency. Although there was a plan, it had never been seriously tested, but the personnel, resources and experience were available. Three Mile Island would thoroughly test the plan and the people.

In spite of this tradition of advance preparation, the Three Mile Island accident occurred at a critical moment in the history of federal planning for nuclear emergencies. On March 30, two days after the accident began, the General Accounting Office published a report to Congress entitled "Areas around Nuclear Facilities should be better prepared for Radiological Emergencies." The report warned that "there is only limited assurance that persons living or working near nuclear facilities would be adequately protected in case of a serious—although unlikely—nuclear accident." The report discussed "deficiencies in planning and preparedness" for emergencies at various facilities operated by the Department of Energy, the Nuclear Regulatory Commission and the Department of Defense. Emergency plans were rarely tested in

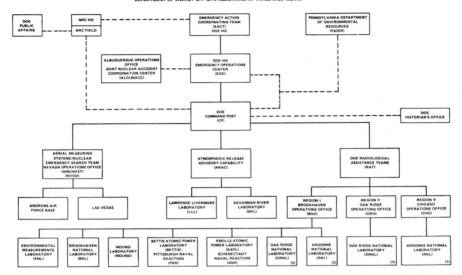

ORGANIZATION CHART
DEPARTMENT OF ENERGY OFF-SITE ASSISTANCE AT THREE MILE ISLAND

(1) ORGANIZATION BEFORE APRIL 7, 1979.
(2) ORGANIZATION AFTER APRIL 7, 1979.

emergency drills, and tests were often limited to the facility itself. Federal agencies and plant operators relied on state and local governments for off-site emergency measures for which they were ill prepared. The report, despite its emphasis on the need for a new federal emergency agency, did not mention the Interagency Radiological Assistance Plan.[38]

Whatever the validity of the General Accounting Office report, its critical tone suggested that the Three Mile Island accident would be, among other things, an opportunity for federal, state and local agencies to show they could respond to a nuclear emergency. Fire or not, the firemen would answer in force.

The Three Mile Island accident also occurred at a time of mounting public concern and debate over the future of nuclear power. The film *The China Syndrome* and the General Accounting Office report reflected this concern. In this atmosphere of doubt, the Nuclear Regulatory Commission withdrew its endorsement of the summary of the Rasmussen Report, the Commission's own study stressing the safety of nuclear reactors. The Commission also shut down five eastern nuclear plants because of their dubious capacity to withstand earthquakes. The Clamshell Alliance, an anti-nuclear group, was campaigning to halt construction of the plant in Seabrook, New Hampshire. The award of damages to the estate of the late Karen Silkwood, a worker in an Oklahoma nuclear plant, suggested that the nuclear industry was not merely negligent, but possibly criminal, in its activities. *The China Syndrome* played upon this suspicion.[39]

These developments were crucial in amplifying the effect the Three Mile Island accident had on both public perceptions and government action. The accident occurred within a political context that questioned not merely the safety of nuclear power in the United States, but its very right to exist. The Interagency Radiological Assistance Plan was unknown to most Americans. To the general public at the end of March 1979, the looming towers of nuclear power plants had become a source of danger, rather than power. Fear stalked the land. Some thought they had seen the fifth horseman of the Apocalypse, and it **was nuclear** power.

Chapter III
The Vanguard

What seemed like a cleanup operation began to grow. It became a political response, not a scientific one.

Robert Friess

T he weather on Long Island was crisp and clear, unusually pleasant for the end of March. Nathaniel Greenhouse, captain of a radiation assistance team at Brookhaven National Laboratory, decided to take advantage of the weather to ride his bicycle the several miles to work. He was finishing his coffee when the phone rang shortly after seven. A security officer at Brookhaven was on the other end of the line. He had just received a call from the control room at the Three Mile Island nuclear power station in Pennsylvania, he told Greenhouse. The caller had reported a possible radiation problem; high levels of radioactivity were being detected in the containment building surrounding the reactor.[1]

Greenhouse noted the control room's phone number and called immediately. Yes, someone told him, the plant had experienced a lack of feedwater and a radiation monitor in the dome of the containment building was reading six hundred Roentgens per hour, enough to be fatal. Possibly, the man added, there was a leak from the primary into the secondary coolant and this was causing the high radiation levels. Was the plant requesting radiological assistance, Greenhouse asked. No, the voice answered, the control room only wanted to alert the radiological assistance team as required by the plant's emergency procedures manual. Then the phone went dead. It was the last contact Brookhaven and the Department of Energy would have with the stricken plant.[2]

Even though the power plant had not requested any help, Greenhouse wanted to be ready if a radiological assistance team were needed. He phoned his department head, Charles Meinhold, and explained the situation. They quickly decided to have all the Brookhaven teams report immediately on a standby alert. What Greenhouse had originally envisioned as an "exercise" was rapidly developing into something much larger. Moreover, his own attempt at exercise, a leisurely bicycle ride to the office, became a casualty of the need for swift preparation.[3]

Within an hour radiation safety experts had gathered at Brookhaven's Safety and Environmental Protection Division to make contingency plans. Meinhold explained that the reactor had suffered some fuel damage. With this in mind, the Brookhaven scientists began selecting the radiation detection equipment they would need, including a newly developed instrument which detected the presence of radioactive iodine. At the same time, Robert Friess, assistant director of the Department of Energy's Brookhaven Area Office, started making arrangements with the U.S. Coast Guard to fly the radiation assistance teams and their equipment to Pennsylvania. Friess had difficulty making the phone call as the Coast Guard's telephone numbers had been changed. He finally reached them and was assured that the Coast Guard would provide two helicopters.[4]

While Friess was trying to line up Coast Guard assistance, his chief, David Schweller, was on another line notifying the Department of Energy's Emergency Operations Center in Germantown of the situation at Three Mile Island. Schweller repeated the information that Greenhouse had received from the Three Mile Island Unit 2 control room. Although the utility was not requesting any assistance, Schweller said, given the high radiation levels reported in containment, he felt the plant had a "substantial" problem. In fact, Schweller told the Emergency Operations Center, conditions warranted sending the Brookhaven radiological assistance team to Three Mile Island. Consequently, he had asked the Coast Guard for support and everyone was standing by to go to Three Mile Island. The large clock on the wall read 8:05. More than four hours had passed since the reactor had tripped.[5]

No sooner had Schweller hung up after reporting to the Emergency Operations Center than his phone rang. Robert Bores, a scientist at the Nuclear Regulatory Commission's Region I headquarters in King of Prussia, Pennsylvania, called to provide current information on the power station. Unit 2 had tripped earlier in the morning, Bores said, and possibly some of the fuel rods had melted. Radiation levels in the containment dome were high, perhaps between two hundred and six hundred Roentgens per hour. Bores also explained that the auxiliary building was contaminated and some radioactive gases had vented into the environment. Had Brookhaven sent anyone to Three Mile Island, Bores asked. Not yet, Schweller answered; did the

Nuclear Regulatory Commisson want a radiological assistance team? Bores said no, he didn't think anyone was needed at this time, but the Brookhaven crew should stand by. By then Schweller believed that, as he said later, something was "very grossly wrong down there."[6]

Brookhaven personnel may have been standing by, but they were not standing still. Convinced that initial reports of the accident indicated a serious problem and puzzled that neither the utility nor the Nuclear Regulatory Commission had requested their assistance, Charles Meinhold, the head of Brookhaven's radiation safety division, sought another way to get the Brookhaven team involved. Meinhold had directed a training program in radiation safety at Brookhaven for many years and had spent over two decades working in the radiation health field. One possible avenue of involvement, he thought, lay through the Pennsylvania Bureau of Radiation Protection in Harrisburg. The director of the Bureau, Thomas Gerusky, had gone to school with Meinhold at the University of Rochester. Later they had worked together at Brookhaven before Gerusky moved to Pennsylvania. Moreover, Margaret Reilly, Gerusky's assistant, was a former student of Meinhold's and had worked with him on a radiation safety project in the Marshall Islands the previous year. So Meinhold called with an offer of assistance. Gerusky, however, was reluctant to initiate a large federal response because there had been no unusual radiation detected off the island. But if anything developed, he promised Meinhold, he would call Brookhaven immediately.[7]

By nine in the morning, five hours after the Unit 2 reactor had malfunctioned and two hours after the utility had notified the radiological assistance team, Brookhaven scientists remained frustrated in their attempt to assist at Three Mile Island. Both Schweller and Friess strongly believed that the team should be assisting. Yet it could not respond without a specific request and Metropolitan Edison had never followed up its initial report or asked for any help. What was occurring on the island after the control room's call to Brookhaven emphasized the developing dilemma: radiation levels were rising within the reactor, but utility monitoring teams were finding only minimal radiation levels off the island. Consequently, neither Bores nor Gerusky cared to initiate a federal response with its attendant costs and public reaction without more conclusive evidence of danger to public health and safety.[8]

☆ ☆ ☆ ☆ ☆ ☆ ☆ ☆ ☆ ☆ ☆ ☆ ☆ ☆

Shortly after seven a.m., the plant had declared a "site emergency" indicating that there had been an uncontrolled release of radioactivity into the immediate plant area. Just minutes after Greenhouse had spoken to the control room, emergency workers were ordered out of the auxiliary building due to rising radiation levels. A Bureau of Radiation Protection engineer who heard the evacuation order being given while he was talking to the control room from Gerusky's office thought to himself, "This is a biggie." Still, the accident appeared controllable.[9]

During the site emergency, control room personnel began to alert state and federal authorities. The Pennsylvania Emergency Management Agency in Harrisburg, the Nuclear Regulatory Commission in King of Prussia, and the radiological assistance team at Brookhaven were all notified according to the long-established emergency plan. Under these emergency procedures this limited notification was all that was required. But the emergency quickly escalated.[10]

Inside the plant the incident certainly looked like a "biggie," and it was getting bigger. An alarm on the radiation monitor in the dome of the containment building had gone off just after the control room talked with Greenhouse. At 7:24 the station manager had declared a general emergency because he felt that potential existed for "serious radiological consequences" to public health and safety. The plant had notified state authorities and the director of the Dauphin County Civil Defense agency. By 7:55, in response to the utility's request, the Nuclear Regulatory Commission opened a direct telephone line between the Unit 2 control room and King of Prussia. No one, however, notified those best prepared to respond to a radiological threat, the team at Brookhaven.[11]

As conditions inside the power plant worsened, off the island the situation appeared stable and safe. Metropolitan Edison sent out survey teams to measure radiation levels at several spots around the plant. While one team reported increasing levels inside the auxiliary building, on the perimeter of the island little radiation was detected. Teams found levels under one millirem per hour on the west shore of the island and the same levels on the east side paralleling Route 441. A Pennsylvania state police helicopter ferried another team across the Susquehanna River to Goldsboro. Again, no unusual levels of radiation could be detected. As a result, the plant operators and Pennsylvania officials concluded that the accident at Three Mile Island was under control.[12]

Nonetheless, radioactivity in the reactor building continued to increase.

Shortly after a group of Nuclear Regulatory Commission inspectors arrived at the plant from King of Prussia, radiation in the Unit 2 control room rose above acceptable levels. The operators strapped on face masks to filter out airborne radioactive particles, but the respirators muffled their voices and communications became jumbled and difficult.[13]

Outside the plant radiation monitoring continued. In the early hours of the accident the state and the Nuclear Regulatory Commission were dependent upon the Metropolitan Edison survey teams for radiation measurements. Lieutenant Governor William Scranton III told reporters that the utility had been monitoring the area around the plant since the accident and "no increase in normal radiation levels" had been detected through mid-morning. Later the Bureau of Radiation Protection sent its own people into the field.[14]

After 10:30 the Metropolitan Edison survey teams reported higher exposure rates, in the 3 millirem per hour range. A Bureau of Radiation Protection crew found similar levels. Quick field estimates by the utility indicated high and possibly dangerous concentrations of radioactive iodine. This discovery, which later proved incorrect, alarmed Maggie Reilly and she decided that the Bureau would have to take a close look at the milk from local dairies. It may also have softened her resistance to Brookhaven's offer of aid.[15]

While the Metropolitan Edison survey teams were looking for radioactive iodine, Brookhaven officials continued to offer assistance. By nine a.m. Schweller had told Bores, the acting duty officer at King of Prussia, that he had two radiological assistance teams standing by. He also explained that additional Department of Energy resources were available in the form of an aerial radiation monitoring team located at Andrews Air Force Base near Washington, D.C. An hour later, Schweller's assistant, Bob Friess, asked Bores if he wanted the Andrews group to make an aerial survey of the radiation levels in the atmosphere above the power station. Bores declined the offer, because no off-site releases had yet been detected. Friess thought that was "crazy" and called the Emergency Operations Center to have the Aerial Measuring System/Nuclear Emergency Search Team (AMS/NEST as they were usually called) put on alert anyway. Friess' decision proved prescient. Moments later Bores called Friess, requesting that AMS/NEST be put on stand-by alert. Bores also wanted to know how long it would take the AMS/NEST aircraft to reach Three Mile Island from Andrews. Friess didn't know but said he would check and get back to Bores. About forty-five minutes later Friess got through the loaded telephone circuits to Bores. The AMS/NEST crew could be at the site within two or three hours of notification. Did Bores also want to make use of the Brookhaven teams, Friess inquired. Again Bores declined, but suggested that Friess re-establish contact with the Bureau of Radiation Protection. Soon, Meinhold was back on the phone to his former student, Maggie Reilly. It was nearly eleven.[16]

☆ ☆ ☆ ☆ ☆ ☆ ☆ ☆ ☆ ☆ ☆ ☆ ☆ ☆ ☆

As the size and seriousness of the accident became more evident during the mid-morning hours, the Emergency Operations Center mobilized to meet the crisis. Located in a series of windowless rooms in the basement of the Department of Energy building in Germantown, Maryland, about twenty-five miles from downtown Washington, the Center kept a twenty-four-hour watch in the event of an emergency. Maps, movie screens, and pictures dotted the walls, telephones were scattered on a long conference table. To one side a photocopying machine stood ready to produce extra copies of documents coming over the telefax system next to it. As people came to work Wednesday morning and word of an accident at a nuclear power plant spread, the sound of concerned voices and jangling telephones quickly began to fill the Center.

The first call had come from Dave Schweller at the Department's Brookhaven area office around 8:45. A call to the Nuclear Regulatory Commission's Incident Response Center in Bethesda, Maryland, confirmed the seriousness of the problem. Still, the Department's resources had not been called in, only notified. During the next hour the staff contacted officials from the Office of Military Application, which was responsible for the Center's day-to-day operation. They also notified officials in the Office of Environmental Safety and the Office of Energy Technology, both of which had an important stake in the events at Three Mile Island. On Friess' recommendation the Center also alerted Herbert F. Hahn, the Department's representative at the AMS/NEST operation at Andrews. At 10:45 Major General Joseph K. Bratton, the Director of the Office of Military Application and the head of the Emergency Action Coordinating Team, ordered that the team convene to respond to the accident as effectively as possible. By the time the team met at eleven o'clock, the Center had already alerted radiological assistance teams from the Bettis Atomic Power Laboratory near Pittsburgh and the Oak Ridge National Laboratory in Tennessee, in addition to those at Brookhaven and Andrews.[17]

Under the Radiation Assistance Plan for Region I, where the Three Mile Island reactor was located, the Department of Energy's Brookhaven office had primary responsibility for any nuclear accident. As events unfolded during the morning, Schweller and Friess were still trying to coordinate the Department's response with the Nuclear Regulatory Commission, Pennsylvania authorities, the Brookhaven National Laboratory, the AMS/NEST crew at Andrews Air Force Base, and finally the Emergency Operations Center. But the complexity of events soon transferred that leadership from the Brookhaven Area Office to the Emergency Operations Center.

Just as the Emergency Action Coordinating Team meeting convened at eleven a.m., the plant supervisor at the Unit 2 control room on Three Mile Island became alarmed at the increasing levels of radiation and ordered all non-essential personnel off the island. The Nuclear Regulatory Commission, which still maintained a direct telephone line to the control room, must have realized the implications of the order and called the Emergency Operations Center with a formal request for the AMS/NEST aerial monitoring team. At the same time, Meinhold was trying to convince Reilly and the Bureau of Radiation Protection that Brookhaven scientists and equipment ought to be at Three Mile Island. The utility thought it had detected radioactive iodine, and Brookhaven scientists had developed just the instrument to monitor it accurately, Meinhold reasoned. Unaware that the field reports on the iodine were erroneous, Reilly finally gave in. "All right, Charley," she sighed, "why don't you come." Meinhold later recalled thinking that Reilly meant only him when she said "you." But he interpreted the word as he wanted to interpret it and made final arrangements to send down the two Brookhaven teams.[18]

The Nuclear Regulatory Commission made it quite clear that it was requesting AMS/NEST assistance and not the resources from Brookhaven. Pennsylvania's Bureau of Radiation Protection, on the other hand, asked only for Brookhaven help. As the need for centralized coordination became clearer, the Emergency Operations Center became the logical answer. Department officials perceived the Three Mile Island situation as an environmental problem, not a nuclear one. Consequently, Whittie J. (Jack) McCool, a long time environmental specialist and Department employee whose career stretched back to the early days of the Atomic Energy Commission, directed the Department's various response efforts from the Germantown Emergency Operations Center.[19] Since the Department of Energy would be responding to separate requests — one federal, the other state — coordination between the field and headquarters would be absolutely essential as the response grew.

The Department of Energy coordinated its people in the field, but did not fully coordinate its activities at headquarters. In this sense, the right hand did not always know what the left was doing. Herbert Feinroth, Chief of the Reactor Evaluation Branch in the Office of Energy Technology, received a message to come to the office of the director of Nuclear Energy Programs, Robert Ferguson. The Assistant Secretary of Energy Technology, John M. Deutch, had heard about the accident at Three Mile Island and wanted a personal representative on the scene, and Feinroth had been selected, Ferguson said. Feinroth agreed and left to hunt up a government car and a radiation badge. There was the inevitable delay in commandeering an automobile and it took two hours to get one. Then word came from Deutch's office: the trip was off. Deutch, unaware of the Department's response capability, had learned that the Nuclear Regulatory Commission was responding to the accident, decided that the problem was not one for the Department of Energy at that time and concluded that Feinroth did not need to go to Pennsylvania. This would not be the last time that the nuclear reactor people proved uninformed about the activities of those in environmental safety.[20]

As in any crisis, many individuals central to the Department of Energy's emergency response were involved in their daily routine when first notified of the Three Mile Island accident. L. Joe Deal, the chief of Environmental Protection and Safety, was attending a meeting at the National Bureau of Standards building in Gaithersburg, Maryland, a few miles from Germantown. Deal had held the meeting there to escape the frequent interruptions likely to occur at Germantown. On the morning of March 28, Deal's careful plans for a quiet meeting soon went awry.[21]

Joe Deal had worked in the radiation health physics field since 1944, first at Oak Ridge after graduation from Lenoir-Rhyne College, then at Brookhaven, and subsequently with the U.S. Navy during several nuclear weapons tests in the Pacific. He had come back to Washington with the Atomic Energy Commission's Division of Biology and Medicine, which set up the first radiological assistance teams. Later, Deal had worked with F. Raymond Zintz, drawing up the Interagency Radiological Assistance Plan intended to coordinate federal efforts during a nuclear accident. As chief of the Environmental Protection and Public Safety branch, Deal continued to have an influential voice in the field of radiation safety. So it was not surprising that the Nuclear Regulatory Commission's Incident Response Center tracked Deal to his hideaway to inform him of the problem at Three Mile Island.[22]

Deal asked his assistant, L.J. Beaufait, to alert the AMS/NEST group at Andrews and the Atmospheric Release Advisory Capability, known as ARAC, at the Lawrence Livermore Laboratory in California. Beaufait left the meeting and attempted first to phone the AMS/NEST headquarters in Las Vegas, but the line was busy. He had better luck reaching Herb Hahn at Andrews. Beaufait's news was the first Hahn had heard about the reactor and he immediately alerted the EG&G scientist in charge of the equipment to get ready. Beaufait tried Las Vegas again. He reached the manager of the AMS/NEST operations, John F. (Jack) Doyle, and told him to prepare for a response to the accident. It was 9:45 a.m. in Washington and three hours earlier in Nevada when Doyle began to assemble AMS/NEST resources. Beaufait then headed back to his Germantown office. From there he called the Livermore Laboratory and asked them to activate the ARAC computers. Within hours ARAC was projecting radiation levels in the plume spreading out from Three Mile Island. These predictions, based on weather conditions, topography and radiological information, were immediately passed on to the Emergency Operations Center and the Nuclear Regulatory Commission.[23]

The initial alert at Andrews produced what may best be described as the "telephone effect." In that children's game a phrase or sentence is whispered from one player to the next down a line or around a circle. The final message is then compared with the original, which, to the players' amusement, is vastly different. The same effect occurred when G. Robert Shipman, the EG&G scientist and acting duty officer, received a garbled message instructing him to prepare the AMS/NEST resources for a possible response to a nuclear accident. Fortunately, the confusion did not affect Shipman's ability to respond. The game began, quite unintentionally, when Shipman phoned the Emergency Operations Center to get additional information shortly after Hahn had relayed Beaufait's alert. He needed to obtain maps of the area around the reactor to plan the monitoring flights. The accident was taking place at the Nine Mile Island reactor on Long Island, Shipman was told. Checking the list of operating reactors, he discovered that there was no Nine Mile Island. But two sounded as if they might be the one — Three Mile Island near Harrisburg, Pennsylvania, and Nine Mile Point on Lake Erie in upstate New York. Realizing that accurate topographical maps would be essential, Shipman dispatched someone to buy from the U.S. Geological Survey maps showing the area around both plants. Although in the end the "telephone effect" did nothing more than stick Shipman with a good number of extra maps of Nine Mile Point, the maps he brought to Three Mile Island proved extremely important, since no other monitoring group or agency had any maps except some hastily obtained at a local gas station.[24]

Part of the message to Shipman had come through very clearly. There had been a "significant release off site," and Shipman, a Ph.D. in nuclear physics from the University of Florida, realized that after operating for a time, a nuclear reactor would build up a sizeable inventory of radioactive fission products such as iodine and cesium. The potential threat, he reasoned, could be "extreme." He immediately went to the AMS/NEST equipment room to check out and pack up the instruments an advance party might use. Since information on the reactor was spotty, Shipman decided to take nearly every type of instrument in the AMS/NEST inventory. Better to haul along too much than too little, he thought.[25]

As Shipman readied the equipment, Herb Hahn wrestled with another problem: how to get the scientists and their instruments to Three Mile Island should it prove necessary. Most of the AMS/NEST aircraft were away from Andrews on another mission and the small H-500 helicopter sat alone in the EG&G hangar. It would be able to ferry some of the equipment or some of the personnel, but it could not carry both. The solution to Hahn's dilemma sat on the runway next to the EG&G hangar — several large Air Force helicopters of the 1st Helicopter Squadron. Hahn phoned the squadron's commander, Lieutenant Colonel John L. Wells, explained the situation and asked if Wells could help out.[26]

Wells recognized the urgency of the request but had no official authorization to order his helicopters on such a mission. Yet time was crucial. Delays could be dangerous. So Wells scheduled a "training flight" to the Harrisburg area and asked for volunteers. He carefully explained that those who might go would encounter "hazards of unknown proportions." A full crew had volunteered in minutes and when the call came to fly, they just "went ahead and went," as one crew member put it.[27]

The request that the AMS/NEST crew assist at Three Mile Island came from Bob Bores of the Nuclear Regulatory Commission shortly after eleven a.m. The aircraft was to fly to Capital City Airport in New Cumberland, about eight miles northwest of the plant, and await further instructions.[28]

Once AMS/NEST had been called to the scene other logistics had to be considered. While Shipman and Ike Harris loaded the advance party's gear into an Air Force helicopter, Hahn arranged that the H-500, owned by the Department of Energy and flown by EG&G, would bring the balance of the equipment later. For Hahn the important task was to establish a command post at Capital City Airport. He had learned in field exercises and earlier call-outs that a coordination center at the scene was essential. Achieving coordination and cooperation among Brookhaven teams working for the state, the AMS/NEST group flying for the Nuclear Regulatory Commission, and the Department of Energy's Emergency Operations Center would not be easy. Moreover, hangar space and gasoline supplies for aircraft, an area for scientists to locate their instruments and field laboratories, a telephone communications center and motel rooms for people working out of the command post had to be arranged. While the scientists and technicians monitored radiation levels, Hahn faced a less romantic but equally vital task. As the first helicopter lifted off and headed toward Pennsylvania, Hahn looked at his watch. It read ten minutes to one. Nearly nine hours had elapsed since the Unit 2 reactor had tripped.[29]

The Capital City Airport in New Cumberland had once been the hub of air transportation for Harrisburg and the surrounding area. But Capital City had become a quiet backwater when Olmstead Air Force Base closed down and was converted to civilian purposes. Renamed the Harrisburg International Airport, the old base had subsequently absorbed the bulk of commercial air traffic. The Governor's plane remained at Capital City and the State Department of Transportation maintained a hangar there. The rest of the airport was used by private planes and the Civil Air Patrol.

The thrupp-thrupp-thrupp of the Air Force helicopter jarred the little airport at 1:45 p.m. on March 28. For the next two weeks Capital City Airport would be the buzzing center of the Department of Energy's response to Three Mile Island.[30]

The advance AMS/NEST party from Andrews Air Force Base touched down and Hahn bounded out of the helicopter to find the airport manager and establish contact with the Emergency Operations Center. Meanwhile, Shipman, Harris and the helicopter crew unloaded the equipment. Dropping out of the blue, as he literally did, Hahn's presence and his insistence on obtaining office space might have caused more alarm than cooperation. But a quick explanation of the situation at Three Mile Island and of the Department of Energy's responsibilities and needs eased any doubts. Moments later Hahn moved into a vacant office, checked in with the Emergency Operations Center, and spoke with the telephone company about installing additional lines. Then he worked on motel arrangements for the AMS/NEST and Brookhaven teams, an activity which foreshadowed a prolonged monitoring effort.[31]

The afternoon of the 28th was clear and cold. The flight from Andrews to Capital City had been uneventful except for one incident. As the helicopter approached Three Mile Island the red light of the Inflight Blade Inspection System (IBIS) blinked on. The system is a safety feature: when lit, it warns the pilot that at least one of the helicopter's rotor blades has been damaged and the aircraft should land immediately. But this time the light blinked off. The pilot assumed that the system had malfunctioned and he continued to Capital City.[32]

Minutes later the Coast Guard helicopter carrying the Brookhaven team arrived. The Bureau of Radiation Protection had sent them to the helipad at the Holy Spirit Hospital, not far from the Bureau's offices in downtown Harrisburg, but the landing area was too small, so the helicopter flew to Capital City. Near the airport, the helicopter narrowly avoided colliding with a small private plane, but the Coast Guard pilot, like his Air Force colleague, was concerned because the light on his IBIS system had lit up. He was also puzzled when it winked off. On the ground, a comparison of notes solved the riddle. Radiation, the two pilots realized, triggered the blade inspection system. Should something happen to a rotor blade, a small radioactive source was released, setting off the warning light on the helicopter's instrument panel. Nothing was wrong with the blades, the pilots and scientists decided. Rather, both helicopters had flown through the invisible radioactive cloud, or plume, as they passed near Three Mile Island. Radioactive xenon escaping from the reactor had triggered the alarm. Since xenon is a noble gas and does not combine with other elements, when the helicopters left the plume, the radioactivity dissipated and the light went out.[33]

While the Brookhaven radiological assistance team did not consider xenon to be an environmental hazard, they were concerned about the possible presence of radioactive iodine. Immediately they "swept" the Coast Guard helicopter with their instruments and began taking soil, air, water and vegetation samples around the airport. All were free of abnormal radioactivity.[34]

By late afternoon cars from the Bureau of Radiation Protection picked up the Brookhaven team and drove them to Gerusky's office in the Fulton Building for a briefing with Maggie Reilly. The office was a chaos of ringing telephones and insistent people. Bob Friess was shocked by the disorganization. Gerusky's staff was simply too small and too harried to cope adequately with the increased demands which had been placed on them. Friess decided that someone from Brookhaven should man a phone in the Bureau's office to facilitate communications with Hahn at the Capital City command post. By then it was apparent to Freiss that his group would be staying overnight. They had not expected that, and none had brought a change of clothing.[35]

☆ ☆ ☆ ☆ ☆ ☆ ☆ ☆ ☆ ☆ ☆ ☆ ☆ ☆

The IBIS alarms on the two helicopters indicated only that there was radioactivity in the air over Three Mile Island. How much radioactivity and where it was drifting were key pieces to the puzzle for the scientists from AMS/NEST. When the rest of their equipment arrived with pilot Jac Watson just before three in the afternoon, Bob Shipman set out to find some of the answers.

Shipman and Watson eased their chopper off the Capital City runway and headed out over the Susquehanna. Shipman wanted to get a "water count" for his instruments. Since water does not conduct radioactivity, Shipman could obtain an accurate background reading by which he could set his highly sensitive instruments. Because the equipment was calibrated to detect very low levels of radioactivity, Shipman wanted to be certain of background radiation in the area in order to determine the levels in the plume.[36]

When the water count was completed, the helicopter began a slow circle toward the plant. It was about two thirds of the way through the arc when Shipman's sensitive instruments started going crazy, the needles on the meters pegged to the limit. Shipman shouted to Watson and the pilot veered quickly away from the area. The needles returned to normal.[37]

Shipman knew that his equipment was sensitive to extremely low levels of radiation, but the fact that the plume from the power plant had saturated the instruments alarmed him. He estimated the level three miles from the plant to be "a few" milliroentgens per hour, yet he could not be certain. Then the two men began to fly in and out of the invisible plume, tracking it with radiation counters. When the count dramatically increased, they knew they had entered the plume. When the count dropped off, they had left it. In this manner they tracked the plume to a ridge line sixteen miles north of Harrisburg. After an hour of tracking, Watson headed back to Capital City. Once more he flew over the river. Shipman's instruments registered normal background radiation levels, which indicated that the helicopter had not been contaminated. This meant, Shipman reasoned, that the plume consisted only of noble gases, primarily xenon, and not the more dangerous fission products such as cesium, iodine or strontium that would have adhered to the aircraft.[38]

As maps, equipment and people overflowed the vacant office Hahn had commandeered earlier in the afternoon, he realized that the growing operation would need more space. The local telephone company asked how many phones should be installed. Off the top of his head Hahn said a dozen. Fine — the installers would be at the airport at eight the next morning. Where should they put the phones? Hahn paused. He would let them know tomorrow when they arrived. He would have to find a new command post.[39]

That evening Hahn checked over three possible command post areas at the airport. From past operations he knew that an area that was easily protected from the public and could be expanded as the operation grew would be best. He selected an unused office area in the barn-red Pennsylvania Department of Transportation hangar. The rooms were dusty and lacked any furniture, but there was more than enough space. The hangar area to the rear of the office space was clean and could also be used. Over the next few days, charts, maps, pictures and signs were taped and tacked to the faded grey-green walls, and picnic tables and chairs supplied by the Middletown Fire Department held stacks of papers, ringing telephones, busy scientists, portable laboratory equipment and a host of people coordinating the activities of the Department of Energy's response at Three Mile Island.[40]

☆ ☆ ☆ ☆ ☆ ☆ ☆ ☆ ☆ ☆ ☆ ☆ ☆ ☆

A flurry of Department of Energy activity was also taking place that afternoon away from Capital City Airport. Personnel at the Emergency Operations Center in Germantown, by then realizing that the accident was a major event, decided that more resources should be transported to Three Mile Island. But before this could be done, lines of command had to be established. There was some confusion over the relationship among different agencies and groups responding to the emergency. No one knew for sure who was running the show for the Nuclear Regulatory Commission. Was it someone from the Region I office at King of Prussia? Several people from that office had been in the Unit 2 control room since mid-morning. Or was it an official at the Incident Response Center in Bethesda? With whom should the command post deal when the AMS/NEST scientists had completed a plume flight? Should someone call King of Prussia, or Bethesda, or Germantown? Pennsylvania officials were also uncertain of the lines of command.

To improve coordination among the Brookhaven radiological assistance teams, the AMS/NEST people, the Nuclear Regulatory Commission and the Pennsylvania Bureau of Radiation Protection, the directors of the Emergency Operations Center decided to send David E. (Ed) Patterson, chief of the Environmental Compliance Office's safety analysis branch, to Capital City Airport as the Department's senior representative at the scene. Patterson signed out a government car and drove to Pennsylvania that evening. With a coordinator at the command post, the Emergency Operations Center staff felt more comfortable about calling in additional resources. Brookhaven agreed to send five additional people and the Bettis Atomic Power Laboratory prepared to dispatch two additional radiological assistance teams of six scientists and technicians each. Another team at the Argonne National Laboratory west of Chicago prepared to come.[41]

The two Brookhaven monitoring teams already on the scene met with Reilly before starting out on their sampling missions. The first team left the Fulton Building in a chilly rain and drove out of Harrisburg toward Hummelstown. Using that day's *New York Times* weather maps to check the prevailing wind direction, the team found the edge of the radioactive plume extending ten miles north of the plant near Hummelstown on Route 441. From there they drove south on 441 to Route 283 north of Middletown, collecting samples of vegetation, soil and water as they went. A second team went out after dinner, around eight o'clock. Samples taken by both groups were analyzed in the Bureau of Radiation Protection lab but nothing unusual was found, notably no radioactive iodine.[42]

Bob Friess conveyed this good news and the results of two AMS/NEST plume flights to a late-night briefing for Pennsylvania Governor Richard Thornburgh and Lieutenant Governor Scranton at the Governor's Mansion. Friess, who had had nothing to eat that day except a piece of cold chicken on the helicopter ride down from Long Island, was grateful for the sandwich Mrs. Thornburgh made for him. The Governor and his staff were even more grateful for Friess' information. The AMS/NEST aircraft was monitoring the radio-active plume, Friess explained, and the Brookhaven monitoring teams were taking soil, plant and water samples. Thus far, he continued, neither group had identified any above-normal levels of radioactivity except from noble gases. They had not found any iodine 131, which would have been a significant danger to public health and safety. In short, there was no danger. Scranton, Friess later remembered, seemed relieved by the information. Exhausted, Friess left the mansion and headed for his motel to sleep.[43]

By the end of the first day nearly everyone at the Capital City Airport command post recognized the need for fresh faces. The morning alert, the excitement, the travel and the continuing monitoring efforts once on the scene had been demanding. Like Friess, everyone was drained. The Brookhaven team had been active since seven that morning. Jac Watson and Bob Shipman had been either flying or preparing to fly since ten o'clock. When the Nuclear Regulatory Commission requested a post-midnight plume flight, Watson begged off. The weather had closed in, visibility was minimal and he was exhausted, Watson told them. The risks of flying under those conditions were simply too high. Accordingly, Hahn convinced the members of the AMS/NEST operation who had just returned to Washington to fly up that evening and pinch-hit for Watson and Shipman the next morning. Another AMS/NEST contingent from Nevada was also en route. So was a fresh radiation assistance team and additional equipment from Brookhaven.[44]

Although it was evident that a large Department of Energy response was underway at Three Mile Island, the radiation levels detected during the first day had not been disturbing. Members of the Brookhaven team, used to responding to an emergency, spotting the radiological danger and then assuring the public's safety, had found no radiological danger and began to wonder why they were needed any longer. Even though sensitive AMS/NEST equipment had become saturated by radioactivity in the plume, concentrations of noble gases, primarily xenon 133, were relatively low. More important, as Friess had told the Governor, they had detected no iodine 131. When tests from samples taken in Goldsboro confirmed to the Bureau of Radiation Protection that no iodine was present, Maggie Reilly became convinced that there were no environmental problems off the island.[45]

Ignored in the confusion and excitement of the first day was the fact that the technical response to the nuclear accident was only a part of the whole. As newspaper reporters and television crews descended upon Middletown and Walter Cronkite reported on the evening news that events at Three Mile Island were "the first step in a nuclear nightmare," the response became political as well as technical. The dichotomy between political and scientific considerations was to lead to misunderstanding on the part of politicians and scientists alike during the accident. Moreover, the technical people had their own confusions and limitations to overcome. For example, the Brookhaven monitoring teams would not have had to rely on weather maps from the *New York Times* had they known about the ARAC plume predictions or checked plume data gathered by the AMS/NEST flights. Radiation assistance had existed for years, but never had so many people been working in one spot before. The scientists from Brookhaven, AMS/NEST, ARAC, and other Department of Energy laboratories would have to learn more about each other's capabilities and then mesh them effectively.[45]

☆ ☆ ☆ ☆ ☆ ☆ ☆ ☆ ☆ ☆ ☆ ☆

Throughout the early hours of Thursday morning, replacements streamed into Capital City Airport. An AMS/NEST computer van, a search van, portable analyzers and other equipment came from Andrews. Five additional AMS/NEST scientists and pilots arrived in a second helicopter to relieve Shipman and Watson, bringing the total of Andrews-based AMS/NEST people to thirteen. Meanwhile, Hahn, who had stayed up all night, began directing the move to the new command post in the Commonwealth hangar. New telephone lines were installed and the fire department moved in tables and chairs. Working with the chief of the airport police, Hahn established tighter security procedures to keep the curious out of the area.[47]

On Thursday, Department of Energy radiation teams continued to monitor off-site radiation levels. After a few hours' sleep, the Brookhaven teams, rumpled in Wednesday's clothing, set out early in the morning to look for radioactive iodine. They found none and felt that their efforts were "almost a clean-up operation." Bob Friess continued to see little threat to public safety and thought the continued presence of the Brookhaven team unnecessary. The teams standing by at the Bettis Laboratory were asked to assume Brookhaven's responsibilities and Friess began to make plans to get his men back to Long Island. The AMS/NEST plume flight that morning confirmed Friess' judgment that there was little public danger. The helicopter detected only low levels of radioactive noble gases thought to be escaping from the auxiliary building.[48]

Since Wednesday, the ARAC computer at the Lawrence Livermore Laboratory had been using weather information to predict the movement of the plume. On Thursday, Marv Dickerson, a Livermore scientist who had helped develop the ARAC system, called Ed Patterson at the command post. He offered to come to Capital City and establish an on-the-spot ARAC operation. In fact, he was over half-way to Harrisburg already, because he had been working in Chicago. Patterson told him to continue east and by the next morning the Capital City command post had established a direct phone link with the ARAC computer at Livermore on an around-the-clock basis. From that time the ARAC system provided meteorological forecasts, predicted plume concentrations and performed dose assessments for plume exposure. This information was then relayed to the Emergency Operations Center and the Nuclear Regulatory Commission's Incident Response Center.[49]

As additional personnel and equipment made their way to Three Mile Island, the Emergency Operations Center also expanded its activities. The atmosphere in the operations room was one of "calm confusion," according to one participant. The event was turning out to be much larger than those in charge had originally anticipated. More resources were arriving and coordination was growing more complex. Others, outside the Department of Energy as well as senior officials within the Department, were clamoring for additional information. Numerous briefings were held for government officials from the Department of Defense, congressmen and their staffs and other concerned parties. The Japanese government wanted the latest information to transmit to its citizens. And while this was going on, the director, Jack McCool, placed additional resources at the Oak Ridge National Laboratory on alert. By mid-morning, the Emergency Action and Coordinating Team had decided to send Joe Deal to Capital City to replace Ed Patterson, with the idea that people should be relieved before they were exhausted. However, soon after Deal arrived at the command post he realized that such a policy would sacrifice continuity. He remained for the next six days.[50]

☆ ☆ ☆ ☆ ☆ ☆ ☆ ☆ ☆ ☆ ☆ ☆ ☆ ☆ ☆

The initial sense of urgency had dissipated by Thursday afternoon. Information on the reactor indicated high levels of radioactivity in the auxiliary building, but only low levels off site. Near the plant the plume still saturated the highly sensitive AMS/NEST equipment and would continue to do so until less sensitive instruments were taken on plume flights.[51]

Communications and coordination between the plant and the Department of Energy command post remained limited at best. Shortly after noon, the Nuclear Regulatory Commission issued its first press release on the accident. There were high levels of radiation in containment and a "continuing release to the atmosphere of detectable levels of radioactive gases." Nonetheless, the release added, the plume measurements were "far below" the thousand millirem level at which the Environmental Protection Agency recommended protective action. Clearly, this information was based on the AMS/NEST monitoring flights. Yet even as Lieutenant Governor Scranton was touring the control room and auxiliary building in an anti-contamination suit, booties and respirator in the early afternoon, a Metropolitan Edison helicopter flying just above the Unit 2 vent stack measured three thousand millirems per hour on its monitors. The Nuclear Regulatory Commission Incident Response Center in Bethesda received a report of this release but never relayed it to Germantown or to the command post at Capital City Airport. Consequently the Department of Energy's aerial measuring team did not monitor the burst. Over two hours later the AMS/NEST helicopter did measure levels of about ten millirems per hour near the plant, but the scientists had insufficient information to link it to the previous release.[52]

Metropolitan Edison officials had specifically requested that the AMS/NEST helicopter not fly over the plant, fearing the consequences of a crash, a company spokesman explained. Thus, only the utility's helicopters monitored the area where the radiation levels were highest. Any other readings timed with a release from the plant had to be made away from the plant and could not be checked against any simultaneous measurements taken over the reactor building. Moreover, either the utility or the Nuclear Regulatory Commission had to inform the command post before a release was expected so that a monitoring flight would be prepared. On Thursday, neither the Nuclear Regulatory Commission nor Metropolitan Edison mentioned that a release would occur.[53]

By Thursday evening the situation around Three Mile Island was characterized by crowding and confusion. When Joe Deal arrived he found conditions at the command post under control, but he soon discovered that the same was not true at the Bureau of Radiation Protection or with the Nuclear Regulatory Commisson staff. At the Fulton Building, he noted that the Bureau of Radiation Protection was overwhelmed, its employees constantly harassed. Local reporters, the national press, ringing telephones, long hours, constant pressure and no relief had "overloaded the circuits," Deal later recalled. Press demands were interfering with vital work. In a quick visit to the Nuclear Regulatory Commission staff located in their makeshift "Trailer City" across from Three Mile Island behind the Metropolitan Edison Observation Center, Deal found a similar situation. Disorganization reigned and Deal had to drag people away from an aggressive television interviewer to discuss coordinating the radiation monitoring efforts. Even the air above Three Mile Island had become congested. Press helicopters whirred around the plant, filming it from all angles and hindering the AMS/NEST monitoring surveys. This risky condition ended when the Federal Aviation Administration at Harrisburg International Airport created a "special use airspace" and barred private aircraft from the area.[55]

If the crush of reporters confused the response effort for the Nuclear Regulatory Commission and the Bureau of Radiation Protection, the command post at Capital City Airport was calm. The airport was isolated from the hordes of reporters and cameramen hunting down the latest news. This was just what Department of Energy officials had wanted. From past experience with other radiological responses, such as Operation Morning Light when many of these same individuals had helped track down the pieces of the Soviet Cosmos Satellite that re-entered the earth's atmosphere and crashed in Canada, they had learned that the public relations part of the mission had to be far removed from the scientific role. The airport police and a contingent from the Civil Air Patrol helped cordon off the command post area and screen out reporters and curiosity seekers. Consequently, after the first day of the accident, reporters rarely visited the command post and the people staffing it were not troubled by the press.[56]

In addition, the Department of Energy had made a policy decision not to speak to the press. The Emergency Operations Center carefully explained to those participating in the response that employees of the Department of Energy and its contractors were at Three Mile Island at the request of the state or the Nuclear Regulatory Commission and working for them. Therefore, since it was not a Department of Energy operation, any public information should come from others. It was this reasoning which had caused the cancellation of Herb Feinroth's trip to Three Mile Island as John Deutch's personal representative. Moreover, the radiological information collected by the Department's monitoring teams had to be analyzed fully before being made public, officials in Germantown believed. The public and the Department of Energy had everything to lose and nothing to gain from a misrepresentation or misreading of the data.[57]

The history of the Department of Energy also affected the decision to keep a low profile during the Three Mile Island operation. As a descendant of the Atomic Energy Commission, the Department was highly vulnerable to criticism that it was "pro-nuclear." The Union of Concerned Scientists had accused the Atomic Energy Commission of covering up unfavorable results of nuclear power reactor tests. If the monitoring results showed little or no harmful radioactivity in the environment, as appeared to be the case, and these figures were released by the Department of Energy, it was thought that this might renew attacks on the findings from already suspicious anti-nuclear activists. Thus the monitoring results were passed on to the Nuclear Regulatory Commission and the Bureau of Radiation Protection for public dissemination. The Department would supply a number of people from its Office of Public Affairs to aid the Commission's efforts, but they would remain in the background. Of course, the Nuclear Regulatory Commission, another child of the Atomic Energy Commission, might also be accused of being pro-nuclear and skewing the data for its own ends, but this was not a Department of Energy concern.[58]

One additional consideration may also have influenced the decision to maintain a low profile. The AMS/NEST operation had worked in secrecy until only recently. What the group did and how they did it had been classified and most of the details still were. There was little incentive to change that situation for reporters at Three Mile Island.

☆ ☆ ☆ ☆ ☆ ☆ ☆ ☆ ☆ ☆ ☆ ☆ ☆ ☆

By Thursday evening the situation, confusing as it may have seemed to those at Three Mile Island, appeared more reassuring to those watching the events unfold on their television sets. Walter Cronkite, who had talked about the China Syndrome and the "first steps in a nuclear nightmare" on Wednesday night, was telling his audience that there was "more heat than light in the confusion surrounding the incident." Other national television commentators also felt that the crisis had passed. A press conference held by Charles Gallina of the Nuclear Regulatory Commission and Governor Thornburgh was just as reassuring. The Governor addressed his remarks to the people who lived near Three Mile Island and told them he believed there "was no cause for alarm." Gallina was even more optimistic. Three times in the course of the briefing he told reporters that the danger to people off-site was over.[59]

Thornburgh was somewhat troubled by Gallina's assertion. His concern turned out to be well founded. Deep inside the reactor vessel a different story was unfolding. Results of an analysis of reactor coolant water indicated far more damage to the core than had been thought. Later the Nuclear Regulatory Commission phoned the Governor with this information, adding that now there was a greater possibility for radioactive releases from the plant. Thornburgh began to question the Nuclear Regulatory Commission's reliability. But none of this new information about the status of the plant was passed on to the Department of Energy or, for that matter, to the Bureau of Radiation Protection. The Nuclear Regulatory Commission had received the coolant report by six-thirty p.m. Thursday, but Joe Deal, who was working with people in Gerusky's office until after midnight, never learned of the new danger. At eight a.m. Friday, when he phoned in his report on Thursday's events to the Emergency Operation Center, he said that the "reactor is under control." At the same time the Metropolitan Edison helicopter was reporting a reading of twelve hundred millirems per hour over the Unit 2 vent stack.[60]

Chapter IV
The Invasion

*There are a number of conflicting versions of
every event that seems to occur.*
Governor Richard Thornburgh
12:30 p.m., March 30, 1979

*The level of hysteria increased directly pro-
portional to the square of the distance from
Three Mile Island.*
Gordon McCleod
Pennsylvania Director of Health

On October of 1938, millions of Americans became alarmed about an invasion that never happened. Orson Welles' radio broadcast of H. G. Wells' "The War of the Worlds" on the night of October 30 convinced many that an army of Martians had just invaded the United States. Conditioned by the march of events toward World War II, men and women assumed that a fictional radio broadcast was describing a real event. The result was widespread hysteria, evacuation, and jammed switchboards across the country. "What time," asked one alarmed caller to the *New York Times*, "will it be the end of the World?"[1]

Although the Three Mile Island accident was fact, not fiction, the media played a mixed role in interpreting it to the American public. Radio broadcasts were probably the best means of communicating up-to-date information. Newspapers carried news that was hours out of date. Telephones were essential, but switchboards were often jammed. Television carried to an anxious public a view of the accident that stressed what might have happened, as much as what was actually happening. Ignorant of the language of nuclear power and radiation, conditioned by films such as *The China Syndrome*, the public lacked both the knowledge and the information to distinguish real from imaginary. The Three Mile Island accident they watched on television was not the total fiction of Welles' 1938 broadcast, but a product of imagination and fear, based on reality. For on Friday experts from around the country converged on Three Mile Island in an invasion that heightened, rather than lessened, the sense of crisis.

Friday began in calm and ended in concern. Until then it seemed that the accident had come to an end, with the plant in stable condition, low off-site radiation readings, and no iodine. On Friday the situation changed. Three Mile Island generated circles of public fear that radiated outward via the media. By nightfall Friday there was talk of explosion and meltdown, a massive weekend exodus from the Harrisburg area, and the arrival of additional Nuclear Regulatory Commisssion experts sent by the White House. Why?

Friday appears to have become a turning point in the history of the accident because of two events: a sudden rise in reactor pressure shown by control room instruments Wednesday afternoon which suggested a hydrogen explosion and which only became known to the Nuclear Regulatory Commission Friday; and the deliberate venting of radioactive gases from the plant Friday morning which produced a reading of twelve hundred millirems directly above the stack of the auxiliary building. What made these significant was a series of misunderstandings caused, in part, by problems of communication within various state and federal agencies. Because of confused telephone conversations between people uninformed about the plant's status, officials concluded that the twelve hundred millirem reading was an off-site reading. They also believed that another hydrogen explosion was possible, that the Nuclear Regulatory Commission had ordered evacuation and that a meltdown was conceivable. Garbled communications reported by the media generated a debate over evacuation. Whether or not there were evacuation plans soon became academic. What happened on Friday was not a planned evacuation but a weekend exodus based not on what was actually happening at Three Mile Island but on what government officials and the media imagined might happen. On Friday confused communications created the politics of fear.

Inside the plant Friday, operators were busy taking action to avoid further damage to the reactor core. From midnight on, with Nuclear Regulatory Commission consent, Metropolitan Edison operators were venting radioactive gases from one tank to another in order to control reactor pressure. This pressure was needed to maintain a proper water level in another tank used to store borated water for cooling the core. This venting went through the auxiliary building vent stack. At 7:56 a.m. a technician in the helicopter took a reading of one thousand millirems. Five minutes later the instrument read twelve hundred millirems.[2]

The venting of radioactive gases Friday morning was carried out at the express instructions of James S. Floyd, Supervisor of Operations for Three Mile Island Unit 2. Floyd had been at Babcock & Wilcox headquarters in Lynchburg, Virginia, when the accident began, and he returned to Three Mile Island on Thursday after running simulation tests of the accident on a Babcock & Wilcox computer. At midnight Thursday, Floyd took over as Shift Supervisor of the crippled nuclear reactor.

Floyd feared running out of borated water in the storage tank, the source of the emergency core coolant system. "I had that one source of water," he later testified, "between me and another LOCA (Loss of Coolant Accident)." At 7:10 Friday morning he therefore ordered an operator to vent the tank. "I knew this would lead to a radioactive gas release into the buildings," recalled Floyd. At 8:34 Floyd called the Pennsylvania Emergency Management Agency and spoke with an unidentified official in a conversation that Floyd later characterized as "one of the classic miscommunications of all times." When asked, "Are you ready to evacuate?" Floyd answered, "Yes, we're always ready to evacuate." It was not clear whether evacuation referred simply to the island or to the surrounding area, or whether Floyd's answer recommended evacuation or expressed a general readiness to evacuate. "I was quoted as ordering the evacuation," Floyd later complained, "which is beyond my authority."[3]

A few minutes later the second communications snafu arose in Harrisburg. At 8:40 Kevin J. Molloy, director of the Dauphin County Office of Emergency Preparedness, called Colonel Oran K. Henderson at the Pennsylvania Emergency Management Agency and reported that "somebody from the plant wanted us to get hold of them in a hurry. I forgot who it was." Henderson's communications officer, Carl Keene, reported a phone call in which the caller was "going ape" about the twelve hundred millirem release. Keene told Henderson, "they have got a serious accident at Three Mile Island." Henderson promptly obtained wind and weather data and called Lieutenant Governor Scranton. At 9:25 Henderson then called Molloy back, reported "some type of release" at Three Mile Island, and said "probably very shortly" Molloy would receive a call to "start an evacuation procedure."[4]

Henderson's concern about evacuation was based on yet another confused telephone call from Washington. The Nuclear Regulatory Commission's Incident Response Center, some one hundred miles from Three Mile Island, lacked direct communication either with Governor Thornburgh in Harrisburg or with its own inspectors in the Unit 2 control room. Among those present that morning was Harold E. Collins, Assistant Director for Emergency Preparedness for the Nuclear Regulatory Commission's Office of State Programs. Around nine the Nuclear Regulatory Commission had received news of the twelve hundred millirem release, producing at Bethesda what Collins recalled as a new "atmosphere of significant apprehension." By coincidence the twelve hundred millirem figure had just been considered as a hypothetical possibility. Coincidence spawned urgency.

Bethesda was now thinking about evacuation. Commission officials Edson Case, Roger Mattson and Harold Denton spoke of "moving people." The Commission relayed the twelve hundred millirem reading to the White House. "I think the important thing for evacuation to get ahead of the plume," said Denton, "is to get a start rather than sitting here waiting to die. Even if we can't minimize the individual dose, there might still be a chance to limit the population dose." Collins was instructed to call Pennsylvania officials and recommend preparation for evacuation. At about nine-fifteen he called Henderson.[6]

Collins asked Henderson what he had heard. Henderson replied that he knew of the twelve hundred millirem release but had not been given any instructions. Collins then said:

> It is the opinion of the management people in this Nuclear Regulatory Commission operations center that you should start thinking about evacuation, and it is the recommendation of these people that you start evacuating people out in the direction of the plume.

Both Collins and Henderson later testified that Collins had then recommended a ten-mile evacuation, but in different words. Collins recalled saying:

> It is our recommendation that you evacuate people in the direction of the plume out to ten miles.

According to Henderson, Collins said:

> We are recommending that you evacuate immediately a ten mile evacuation around Three Mile Island.

Was evacuation to begin or not? If so, was it to be along the plume or in a ten-mile circle? Neither man made his intention clear. Henderson added that he could only plan for five miles. About five or ten minutes later Collins called back and said the Commission would support the five-mile figure. Henderson promptly called Scranton and recommended evacuation.[7]

In Harrisburg, state officials were divided over the evacuation recommendation from the Nuclear Regulatory Commission. The Pennsylvania Emergency Management Agency was recommending it on the basis of Collins' phone call and the twelve hundred millirem release from the plant. Lieutenant Governor Scranton later testified that Henderson "called me at home and said Mr. Collins from the Nuclear Regulatory Commission — and I didn't know who Mr. Collins was — said there was an unplanned release of twelve hundred mr from the island and that he thought we had to evacuate." But both Dornsife and Gerusky of the Bureau of Radiation Protection opposed evacuation on the grounds that off-site radiation readings were consistently lower than on Thursday. Gerusky felt that the Nuclear Regulatory Commission recommendation should have been approved by his office rather than directly going to Henderson and Scranton. Collins' telephone conversation was "a wild one" that needed corroboration, he later observed. And since Gerusky did not agree with the recommendation, he sent Dornsife to Thornburgh's office to try to stop the evacuation.

The Bureau of Radiation Protection and other opposition to evacuation was based on more extensive monitoring data than the one reading used in Washington. Charles Gallina was in the Unit 2 control room around ten a.m. and recalled that "everything was going pretty smoothly at the plant." Suddenly a plant worker ran into the control room and reported that his wife had just heard that the Nuclear Regulatory Commission was recommending evacuation. Confused, Gallina promptly called the Region I office. People there had heard of Collins' recommendation, but did not believe it was necessary. Gerusky shared the same view. So did Metropolitan Edison's John Herbein, who told a late-morning press conference at the American Legion Hall in Middletown that he "hadn't heard the number twelve hundred," and that evacuation was unnecessary.[9]

Faced with division among his advisors, Governor Thornburgh decided to call the Nuclear Regulatory Commission. Before he could, Commissioner Hendrie called from Bethesda to cancel Collins' evacuation order and apologize for the "staff error." Admitting that "we are operating almost totally in the blind," Hendrie recommended only "that it would be desirable to suggest that people out in that northeast quadrant within five miles of the plant stay indoors for the next half hour." Paul Critchlow, press aide to Governor Thornburgh, then issued a statement from the Governor advising those within a ten-mile radius of the plant to stay indoors and keep their windows shut.[10]

The evacuation rumors initiated by the Collins' telephone call to the Pennsylvania Emergency Management Agency were shortly reinforced by other events. At about 11:09 there was another release from the plant. A few minutes later an air raid siren went off in downtown Harrisburg, creating considerable confusion. The Middletown telephone system became overloaded, and school buses were readied to evacuate children to Hershey Park. At the same time President Carter called Governor Thornburgh and told him he was sending Harold Denton to Three Mile Island as a special emissary, adding that it was best to "err on the side of safety and caution." By 11:30 some fifteen thousand state office workers were leaving for the weekend, adding to the unplanned exodus of thousands of fearful citizens from the area. Around that time Hendrie recommended to Thornburgh a partial evacuation of the five-mile radius area.[11]

At twelve-thirty Governor Thornburgh, with the morning's events in mind, held a press conference in Harrisburg. Present were Lieutenant Governor Scranton, Tom Gerusky and Craig Williamson, deputy director of the Pennsylvania Emergency Management Agency. After noting that he had spoken with Carter and Hendrie, Thornburgh said that there was "no reason for panic or implementation of emergency measures," but that

> based on advice of the Chairman of the Nuclear Regulatory Commission and in the interest of taking every precaution, I am advising those who may be particularly susceptible to the effects of radiation, that is, pregnant women and preschool age children, to leave the area within a five-mile radius of the Three Mile Island facility until further notice.

Despite this advisory, Thornburgh emphasized that "there is no evacuation being ordered or undertaken."[12]

Thus by Friday afternoon the twelve hundred millirem reading had triggered an evacuation recommendation, later withdrawn, and led to the more moderate advisory of Governor Thornburgh. Washington seemed more concerned than Harrisburg, and certainly than Metropolitan Edison.

The second major factor that heightened concern on Friday was the belated report of a hydrogen explosion Wednesday and the presence of a bubble within the reactor core. At nine-fifty a.m. the Nuclear Regulatory Commission in Washington transmitted its laconic "Preliminary Notification of Event or Unusual Occurrence" which noted that Unit 2 was "continuing to remove decay heat through the steam generator." It also reported two bubbles in the reactor, one in the pressurizer, an intentional means of maintaining pressure level, and "one in the reactor vessel head caused by the accumulation of non-condensible gases from failed fuel and radiolytic decomposition of water."[13] This was the first official admission of fuel damage, and the first mention of the famous "hydrogen bubble."

Regarding the twelve hundred millirem reading, the Nuclear Regulatory Commission notification said only that "at eight-forty a.m., on March 30 the licensee began venting from the gaseous waste tanks. The impact of this operation is not yet known."[14]

What alarmed the Nuclear Regulatory Commission most was that around ten that morning it learned from Metropolitan Edison that there had been a pressure spike on a Unit 2 control room instrument at one-fifty Wednesday afternoon. "We are guessing," Roger Mattson told Hendrie, "that may have been a hydrogen explosion. They [Metropolitan Edison], for some reason never reported it here until this morning. That would have given us a clue hours ago that the thermocouples were right and we had a partially disassembled core."[15] Mattson calculated that there was now a one-thousand-cubic-foot bubble, "mostly hydrogen," in the upper head of the reactor vessel.[16] Commissioner Kennedy mentioned "a chance of having an explosion from the bubble." Mattson admitted that "it is a failure mode that has never been studied. It is just unbelievable." He speculated that "we probably melted some fuel" and had a "severely damaged core." By two o'clock Mattson admitted that "we have got an accident that we have never been designed to accomodate." He warned that "I don't know what you are protecting by not moving people."[17]

On Friday the White House involved itself more directly in the Three Mile Island accident. Colonel William Odum, military aide to National Security Adviser Zbigniew Brzezinski, ordered a helicopter to fly Harold Denton to the plant.[18] Around noon President Carter watched videotapes of news broadcasts from Thursday evening and Friday morning and was reportedly disturbed by exaggerated media reports of the danger. Brzezinski called Jack Watson, the presidential assistant for intergovernmental relations, and assigned him to coordinate federal and state response efforts. Brzezinski characterized the situation as "at best uncertain" and Jessica Matthews, President Carter's National Security Adviser on Nuclear Reactors, warned Watson that the accident "could be very serious.[19]

At one-thirty the White House convened a special meeting to review the situation and coordinate federal action at Three Mile Island. Chaired by Brzezinski, the meeting included Watson, Matthews and Jody Powell of the White House staff, Chairman Hendrie and representatives of the Department of Defense, the Federal Disaster Assistance Administration and the Department of Energy. Hendrie described the plant situation: the reactor was cooling down; core damage was believed to be greater than originally estimated; low-level radiation releases had occurred that morning as a result of procedures to cool down the plant; and Denton was due on site to take charge of federal actions. Hendrie also described the potentially serious bubble problem. After some discussion the White House agreed that Denton would handle technical matters, Powell would deal with the press and two others would be named to handle evacuation planning and liaison with Governor Thornburgh.[20]

By mid-afternoon the scene shifted back to the plant itself. Around two, Harold Denton and twelve colleagues arrived at Three Mile Island. After meeting briefly with Herbein and telephoning the President, Denton established a temporary operations center in a private home and then held a brief press conference in the yard. According to Denton, communications were "just frightfully inadequate," the Nuclear Regulatory Commission was operating out of a borrowed highway patrol trailer and the reactor core was more seriously damaged than had been thought. Denton told Hendrie privately that they were even considering a "possibility for the melt of the core."[21]

In the popular mind the idea of a "meltdown" of the reactor core through overheated fuel elements, the famous China Syndrome, was the ultimate nuclear plant disaster. The movie and several books had portrayed it. But shortly before four Hendrie called Governor Thornburgh and assured him that neither a hydrogen explosion nor a meltdown were likely. Originally, the

Commission announced only "high fuel temperatures" and a "large bubble of noncondensible gases in the top of the reactor vessel."[22] But at a four o'clock newsbriefing in Bethesda a Commission official, Dudley Thompson, used the word "meltdown" and thereby provoked the following memorable press release:

(Meltdown) (By Alice Cuneo)

Washington (UPI) — The Nuclear Regulatory Commission said this afternoon the Three Mile Island atomic power plant in Pennsylvania faces "the ultimate risk of a meltdown" — the most serious type of nuclear catastrophe — within the next few days as engineers try to cool down the crippled reactor.

Dudley Thompson, a senior official in the Nuclear Regulatory Commission Office of Inspection, said the threat was posed by a steam bubble inside the reactor that could increase in size as pressures within the reactor are lowered, leaving the core without vital cooling water.

"We are faced with a decision within a few days, rather than hours (on how to cool down the core)," Thompson told reporters at a Nuclear Regulatory Commission news center.

"We face the ultimate risk of a melt-down" depending on "the manner we cope with the problem. If there is even a small chance of melt-down, we will recommend precautionary evacuations."

The story was immediately disavowed by the Nuclear Regulatory Commission in a six-thirty announcement that "no immediate danger of a core melt" existed. But it was too late. The story was out.

Such misunderstandings and confusions were commonplace by Friday. The local *Harrisburg Patriot* wrote that residents were now "frightened and concerned" and that "the credibility of the experts went out with the radiation." The *New York Times* editorialized that "the profusion of explanations and contradictions has meant troubling confusion;" "the reactor's operators said one thing, state officials another, federal officials yet another, not to mention the contributions of the equipment manufacturers and politicians." The *Washington Post* also noted "separate and conflicting statements." While Metropolitan Edison's Herbein was telling residents they had nothing to fear because radiation was "insignificant" and "minuscule," antinuclear scientists George Wald and Ernest Sternglass were in Harrisburg saying that radiation levels were fifteen times higher than normal and demanding a permanent shutdown of the reactor. Whom was a citizen to believe?[24]

Friday evening Harold Denton became familiar to millions of Americans. His immediate task was to convey an accurate picture of the reactor's threat to public health and safety directly to Governor Thornburgh. In the larger sense Denton had to explain clearly the science and the technology to a confused public. At Middletown, Denton had come to question the wisdom of evacuation which he had proposed from the distance of Bethesda. He realized that garbled communications among exhausted scientists, persistent newsmen and a fearful public had created "a madhouse." He also noted that "the utility is a little shy, in my view, of technical talent." But by 6:45 that evening he could report that:

> The information which I was going to relay to the Governor is I think we are stable, that the releases themselves, even if they were to continue, which they occasionally have, don't warrant evacuations, that I think the condition of the core doesn't warrant evacuation.[25]

From eight-thirty to ten that night, Denton, Thornburgh and Henderson met in Thornburgh's office to discuss the situation. While stressing the "stable" situation at the plant and the unlikelihood of disaster, Denton mentioned that evacuation plans out as far as twenty miles and involving 700,000 people might be prudent. Henderson testified later that Denton also "discussed the hydrogen bubble," and "was talking terms of a core meltdown, however remote."[26] At a ten o'clock press conference, Thornburgh calmly emphasized that "no evacuation order is necessary at this time," that his earlier advisory that people within ten miles of the plant stay indoors would be lifted at midnight, but that his advisory on preschool children and pregnant women remained in effect. Denton also reassured the press that the possibility of a meltdown was "very remote," and that there was no "imminent" danger to the public.[27]

These reassurances came too late to alter the shift from calm concern to confused fear in the media. "Today Show" coverage Friday morning still minimized the danger. But on the evening news the alarm was clear. Frank Reynolds called the accident "much more serious." Citing Thompson's statement, he spoke of "the possibility, though not yet the probability, of what is called a meltdown of the reactor core" which would lead to "a catastrophe." Walter Cronkite philosophized about Prometheus, Frankenstein and man's "tampering with natural forces," adding that "we are faced with the remote, but very real, possibility of a nuclear meltdown at the Three Mile Island atomic power plant."[28] Such alarmist comments only reflected the confusion and contradictions in Thompson's and Denton's statements, and television commentaries magnified the potential for disaster.

Despite official assurances, Friday had been a day of fear. Perhaps 100,000 people had now fled the area. Mayor Kenneth Myers of Goldsboro described it as a ghost town with empty streets. Robert Reid, Mayor of Middletown, ordered a curfew and announced that looters would be shot on sight. Allegheny Airlines cancelled five of its flights into Harrisburg. Nonetheless, most citizens who stayed behind remained calm, if puzzled. And their confusion was understandable, for rumor and misinformation spread far beyond Three Mile Island.

☆ ☆ ☆ ☆ ☆ ☆ ☆ ☆ ☆ ☆ ☆ ☆ ☆ ☆ ☆ ☆

For Department of Energy personnel at Three Mile Island, Friday was also a day of confusion. For two days the Brookhaven radiological assistance team had searched for iodine and found only xenon. Thursday afternoon they went home tired, hungry and dirty, thinking that the emergency was over. Wednesday and Thursday had been a successfully completed exercise, they thought. Tony Greenhouse later said there was "no further need for us to stay." Bob Casey found the situation "precautionary" but not serious. This feeling was reinforced by the fact that Brookhaven personnel had been working with their old friends and fellow scientists Tom Gerusky and Maggie Reilly of the Bureau of Radiation Protection and not the evacuation-oriented Pennsylvania Emergency Management Agency. The Brookhaven team went to bed on Long Island Thursday night believing the crisis was over. It was not.[29]

To the Brookhaven scientists the events of Friday morning seemed far more alarming from the periphery than at the center. Bob Casey heard radio reports about "uncontrolled releases" from the plant and "potential evacuation" of the area. He could not believe Governor Thornburgh's advisory because he thought the situation was not serious enough to warrant such a response. Casey promptly called Gerusky's office and found that, as he put it later, "they were as dumbfounded as we were." Statements by the Nuclear Regulatory Commission and the news media were creating an impression of imminent catastrophe, overestimating the risk to the public and providing contradictory statements. Thornburgh's advisory was a "big mistake," thought Tony Greenhouse, and "evacuation should never even have been considered as a possibility." Bob Casey thought Walter Cronkite's seven o'clock newscast that evening created the impression of a "war zone" at Three Mile Island.[30]

The Brookhaven team's astonishment was of course understandable. They were no longer seeing events with their own eyes, but through those of the government and the media. Since many decisions were coming from Washington, the view from a distance was far more alarming than from close up. They could not believe that the Department of Energy ordered them back that Friday morning. "Every event which led to us going back," as one of them put it, "was a non-event."[31]

The discrepancy between direct scientific data and media reporting increased Friday. That morning Jim Sage called Brookhaven and reported that the Bettis crews found virtually no iodine in samples taken on both sides of the Susquehanna River. But there would be more "planned releases," he added, and Brookhaven would be needed for "several more days." The Nuclear Regulatory Commission now wanted an instrument calibrated to read very high radiation levels without becoming saturated. Brookhaven immediately sent it by car. Were the releases planned or unplanned? Was Joe Deal now planning to bring in the teams from Argonne and Oak Ridge, or to use Brookhaven, or both? By noon Friday the invasion of the Three Mile Island area by new Department of Energy scientists, along with the Nuclear Regulatory Commission and the media was beginning to generate its own sense of confusion.[32]

In the meantime the Department of Energy was continuing its monitoring operations out of Capital City Airport. The soon-to-be famous twelve hundred millirem release which had occurred at eight a.m. had not been monitored; Tom Maguire and Jac Watson made the first plume flight from ten-thirty to eleven-thirty and found what they considered rather low readings — well below thirty millirems within a quarter mile of the plant. However, the command post at the airport was getting no information directly from the plant and by mid-day was hearing rumors of a "nasty hydrogen bubble in the core," and possible evacuation. "The major cause of our fear," recalled Walt Frankhauser later, "was probably just a lack of information; rumors just fly under those sorts of conditions." In addition, it was impossible to know when to make plume flights. In theory, Metropolitan Edison would let the Nuclear Regulatory Commission know of a planned release and the Nuclear Regulatory Commission would notify the Department of Energy, which would conduct a plume flight to check radiation levels. In fact this rarely happened, and most Nuclear Regulatory Commission requests to fly were to monitor radiation releases that never occurred.[33]

On Friday the pilots were at least getting some relief. Pilots from the 1st Helicopter Squadron, who had been ferrying anxious Congressmen and the Nuclear Regulatory Commission officials up from Washington on Thursday and Friday, now helped fly plume flights. These Friday flights were being complicated by a profusion of press helicopters over the island and the grounding of the AMS/NEST H-500 helicopter due to engine trouble.[34]

Three flights on Friday revealed a continuing pattern of low-level releases of noble gases from the plant. Most readings were below ten millirems, and during the day the plume shifted from the southeast to the northwest. Such low-radiation data gave no reason for the growing concern across the nation about events at Three Mile Island.

Friday's major problem for many involved at Three Mile Island was communications. The Department of Energy was no exception. Joe Deal complained that the phones were "overloaded" when he called the Emergency Operations Center at ten-thirty that morning. He reported the eight o'clock twelve hundred millirem release as "fairly hot stuff," but could add nothing more. Henderson had recently called Deal and said that the Nuclear Regulatory Commission was recommending evacuation, but on checking Deal gathered that the Nuclear Regulatory Commission officials at Three Mile Island were angry and "don't know what's happening." McCool told Deal that the Emergency Operations Center was "getting conflicting reports on whether evacuation was ordered or not." "That's what we're trying to find out, too," said Deal.[35]

At five past eleven Deal again talked to McCool at the Emergency Operations Center and reported that "the King of Prussia [Nuclear Regulatory Commission] people tell us that there was a conflicting signal on the evacuation." On Commission advice, he reported, Thornburgh was recommending simply that people near the plant stay indoors.[36]

Because of heavy telephone traffic, they agreed that Deal would simply call in a report from the command post to Germantown every half hour. Throughout Friday, the Department of Energy continued to pursue its main mission: providing monitoring data, ARAC meteorological predictions and other assistance to Pennsylvania and the Nuclear Regulatory Commission as outlined in the Interagency Radiological Assistance Plan. In addition, the problem of people "dipping in and out at random," as the Brookhaven team had done, was solved by agreeing to coordinate all activity through the command post and Department of Energy headquarters. By noon Friday the Argonne and Oak Ridge radiological assistance teams had been activated and were on their way to Pennsylvania.[37]

Early that afternoon both Metropolitan Edison and the Nuclear Regulatory Commission asked for additional Department of Energy support. At twelve-thirty, Joe Lenhart in Oak Ridge called the Emergency Operations Center and reported that Herman Dieckamp, Vice President of General Public Utilities, had requested a special team to monitor iodine and help assess the composition of the hydrogen bubble. The Emergency Operations Center was also trying to soothe Congressional staff members who called to ask about a "meltdown" or "China Syndrome" at Three Mile Island. Gerry Combs, working in the center, told them that the Department of Energy was "not handling the incident" but simply "providing radiological assistance." As Combs informed one caller from the House Science and Technology Committee, the accident was "serious as far as the reactor is concerned; as far as public health and safety, not so bad." The main source of radioactivity, he noted, was xenon, "and the body has no use for xenon. They haven't seen any iodine at all."[38] While continuing its monitoring activities and briefing Congressional staffers, the Department of Energy began one of its most important support efforts, analyzing the hydrogen bubble.

Around one-thirty in the afternoon, Sol Levine of the Nuclear Regulatory Commission telephoned L. J. Ybarrondo, an EG&G scientist at the Department of Energy's Idaho National Engineering Laboratory near Idaho Falls. Could his staff help with the bubble? Ybarrondo thought so and immediately assembled a group of EG&G experts. A few hours later the Nuclear Regulatory Commission asked if Ybarrondo could use the Idaho Laboratory's reactor model to investigate venting the bubble. The Commission also wanted information on the bubble's detonation potential. By late evening Ybarrondo's team recommended that the Nuclear Regulatory Commission continue cooling the reactor. In the meantime Idaho would test a semiscale model of the reactor, using electrical power and nitrogen gas rather than nuclear power and hydrogen, to see if the bubble could be vented. But actual tests, Ybarrondo warned, could not begin until Saturday morning, when a computer named "Puff, the Magic Dragon" would become available.[39]

Meanwhile the Department of Energy became involved with the Friday afternoon White House meeting. Environment and Defense Programs, the two branches directly participating at Three Mile Island, were somehow not represented at the meeting. Energy Technology was. The acting Assistant Secretary for Energy Technology on March 28 was John Deutch. Secretary James Schlesinger had designated Deutch, who had experience in the area of nuclear waste management, as the Department spokesman on Three Mile Island. The Department of Energy traditionally looked to Energy Technology for advice on nuclear energy policy, but not on military or environmental matters. Although it had had no prior experience with the Three Mile Island reactors, Energy Technology became the point of contact with other federal agencies and the White House.

Energy Technology representatives to the White House meeting Friday afternoon were Herb Feinroth and Jack Crawford. Feinroth had had fifteen years' experience in Admiral Rickover's naval reactors program before he joined the Atomic Energy Commission and the Department of Energy. Crawford, a former Rickover assistant, substituted for Deutch, who had received the invitation to the meeting. Feinroth recalled later that there was a "lot of running in and out to be sure that we were getting adequate telephone communications" with both Three Mile Island and Harrisburg. Brzezinski found the inadequate communications to be "unacceptable" and had the Army Signal Corps set up new lines from Washington to the site. After Hendrie's report that the reactor was "stable" but that the hydrogen bubble represented a "potential problem," questions followed about the plant and the possible need for evacuation.

By now evacuation had become a central concern in Washington, if not at Three Mile Island, but no one could reach any final decisions as to how far or how many people would be affected. Hendrie had no answer. Yet the rest of the meeting focused on planning an evacuation without unduly alarming the public. Existing disaster plans seemed inadequate. The Federal Emergency Management Agency existed only on paper. Someone labeled documents on disaster planning as "no good."[40]

For the Department of Energy, the meeting provided little guidance. Neither Feinroth nor Crawford had been involved with the Interagency Radiological Assistance Plan or the Nuclear Emergency Search Team and neither was familiar with the Department's emergency plans and capabilities. All Crawford could offer Hendrie was the Department's assistance in handling waste materials. Later that afternoon Feinroth reported on the meeting to Deutch, and the two men drafted a letter to Secretary Schlesinger. But the Energy Technology link with the White House went no further and other Department operations continued at Germantown and Three Mile Island unaware of and unaltered by the meeting.

By three o'clock that afternoon Bernie Weiss of the Nuclear Regulatory Commission called the Emergency Operations Center with current data. Cooling the reactor, he said, would involve intermittent releases of radioactive gases into the atmosphere. There was evidence of severe fuel damage in the core, and of a large gas bubble.[41] This was the first communication the Department of Energy had received that day depicting the seriousness of the situation at Three Mile Island.

At three-thirty Joe Deal telephoned McCool at the Emergency Operations Center. Denton's staff didn't "know what to do with what they have got. We are going to have a meeting with them here tonight at seven," Deal added, "and we are going to give them a map. They don't even have a map of the area; you won't believe it, they are working on a sketch of a plan and that's it."[42]

By now the command post had enough telephones. The Bettis radiation teams were monitoring off site, and arrangements were being made with the Department of Energy's Lawrence Livermore Laboratory in California to make ARAC's data on weather conditions and radiation readings more available. At Trailer City, where the Nuclear Regulatory Commission had established headquarters, television reporters constantly interfered with the scientists' work. Seeing this, Deal made certain his staff did not talk with reporters or appear on television. The invaders at Capital City Airport were not the media, but scientific experts. Gerry Combs observed to Deal that "people are coming out of the walls," and when the Victorine Instruments Company offered to provide an additional two hundred health physicists, Deal answered, "Oh, God, no."[43]

The Invasion   65

The exodus of citizens was now being matched by an influx of experts.
Once the White House became involved, the Three Mile Island area was over-
run by officials, not only from the Department of Energy and the Nuclear
Regulatory Commission, but from the Department of Defense, the Federal
Disaster Assistance Administration, the Department of Health, Education and
Welfare, the Federal Drug Administration and other federal agencies. The
newcomers, observed Deal, are "all running around and nobody knows what
the hell they got." The Bettis and Brookhaven teams had responded to a request from state officials. Those from Argonne and Oak Ridge had come at
the request of the Nuclear Regulatory Commission as had the AMS/NEST
team. To coordinate and provide continuity to these efforts, Deal offered to
stay through the weekend.[44]

Throughout Friday evening more Department of Energy teams arrived.
Four representatives from the Argonne National Laboratory drove in around
eight, and another five-man team was driving east in a van bringing additional
monitoring equipment. Oak Ridge sent six men that evening and two more
on Saturday with several dozen respirators. Later Friday evening the
Brookhaven team returned to join the battery of experts now laying siege to
the crippled plant.

Meanwhile the Capital City command post was quietly coordinating federal
and state efforts at Three Mile Island, not in response to any specific instruc-
tion, but simply because the Department of Energy had been in the area for
more than two days and had the experience and technical capability to do the
job. Plume flights were flown on a fairly regular basis. Detailed maps, aerial
photographs, and charts covered the command post walls. The telephone
communications at Capital City were superior to those available through the
Nuclear Regulatory Commission or the plant. In addition, Marv Dickerson
had set up a telecopier that provided accurate meteorological predictions directly from a computer at the Livermore Laboratories. Because of its early
presence and technical capabilities, the Department of Energy operation now
became, unknown to the press or the public, the *de facto* center of the effort to
monitor radiation and coordinate state and federal data-gathering efforts.

Friday afternoon Joe Deal had arranged a meeting for that evening at
Capital City Airport to coordinate all environmental monitoring efforts at
Three Mile Island. Deal felt that the isolated location would avoid the confu-
sion which existed at Trailer City and the office of the Bureau of Radiation

Protection in Harrisburg. The meeting began at seven, with state and federal technical experts representing the Department of Energy, the Federal Drug Administration, the Environmental Protection Agency, the Department of Health, Education, and Welfare, and Pennsylvania's Bureau of Radiation Protection present. Nuclear Regulatory Commission representatives arrived about seven-thirty. Each agency briefly summarized its activities.

The Federal Drug Administration representative reported that the state milk analysis facilities had become overloaded, and that the Federal Drug Administration would use its laboratory at Winchester, Massachusetts, to ease the burden. The Environmental Protection Agency promised that air sampling aircraft and ground sampling equipment would arrive from Las Vegas the next day. Joe Deal explained what the Department of Energy was doing. He said that the Brookhaven and Bettis scientists had taken one hundred and fifty samples, analyzed a hundred and ten of them and found no traces of iodine. The Bureau of Radiation Protection reported continuing ground monitoring with dosimeters in Falmouth, Goldsboro, Middletown and close by the reactor. No one present was sure how often the dosimeters were being read or how they were calibrated , but the meeting allowed the uncertainty to surface and enabled the AMS/NEST scientists to iron out any problems. Finally, there were brief reports by Jack Doyle on the AMS/NEST plume flights, Marv Dickerson on the ARAC predictions, and Jim Sage on the results of Bettis radiological assistance team monitoring.[45]

The meeting made it clear how substantial the Department of Energy effort was. There were sixty scientists and support personnel at Three Mile Island by Friday evening, working under the direction of Joe Deal and the Emergency Operations Center. Monitoring data were undergoing analysis, plume flights were proceeding on schedule and wind-and-weather predictions were being carried out. Not surprisingly the various representatives on the spot decided "that one agency ought to be responsible for collecting information on data gathered by all the agencies and since the Department of Energy seemed to have the greatest resources, it was asked that they do this job."[46]

Beyond the command post and those directly involved with the monitoring efforts, the Department's role remained largely unknown. Neither aerial monitoring nor the Interagency Radiological Assistance Plan was mentioned at the Friday White House meeting or in the press. The Nuclear Regulatory Commission knew of the flights, of course, and may even have thought that the Department's aircraft, rather than the Metropolitan Edison helicopter, were the source of the twelve hundred millirem reading. Harold Denton testified later that "we went back to the national labs and the Department of Energy for a lot of this expertise." In fact, Friday's events illustrated the existence of two levels of federal involvement at Three Mile Island: The White House-Nuclear Regulatory Commission coordinated response that began Friday afternoon, and the Department of Energy-Interagency Radiological Assistance Plan operation that had begun two days earlier when AMS/NEST and Brookhaven teams arrived on Wednesday. The White House stood in the national and international spotlight, and the Department's efforts remained largely unknown. "I was not aware of the magnitude of the massive effort that the Department of Energy had provided," recalled John Villforth, a Department of Health, Education, and Welfare scientist, "until later when I had gotten up there. I don't think in the very beginning that I appreciated that this was, in fact, an Interagency Radiological Assistance Plan exercise and that the teams were there under this Plan."[47]

By Friday evening most of the area around Three Mile Island was occupied by an army of federal and state experts who faced an opposing army of questioning reporters. In the search for news, reporters interviewed other reporters. Many civilians in the battle area had fled. A survey of area residents revealed that fifty-three per cent of those living within twelve miles of the Three Mile Island plant had evacuated some or all of the members of their family, compared with only nine per cent of those beyond twelve miles. Over half of those who did evacuate the area at all did so on Friday, and most stayed with friend or relatives in other parts of Pennsylvania. None of this happened according to plan, and evacuation routes appeared in area newspaper only on Monday. By late Friday many occupants of Middletown and Goldsboro were not residents but scientists and newsmen.[48]

On Friday Three Mile Island became the focus of world attention in the news media. Fear, confusion and uncertainty radiated outward, initiated by contradictory statements from government experts and amplified by the press. Television brought the crippled plant into American living rooms. The further away from the center, the greater the fear. A Chinese report said that eight million people had been evacuated from around Three Mile Island, while Amish folk in nearby Lancaster County's Pennsylvania Dutch region, who had no electricity or television, did not know there had been an accident.[49] Nuclear Regulatory Commission officials in Washington speculated about a possible meltdown, while Commission experts inside the Unit 2 control room were surprised to learn that there had been an evacuation advisory. No wonder many nearby residents who watched the news that night drove off in the family car for the weekend.

The fact that Three Mile Island became a media event on Friday did not mean that the danger was not real. Many scientists and nuclear industry experts were deeply concerned about the reactor. But accurate information was in short supply because of poor communications, and public fear increased as media reports became more alarming. An invasion of experts and an exodus of residents was the result. Having arrived as the vanguard, the Department of Energy had established the beachhead on what was about to become a battlefield.

Entrance sign to Three Mile Island on Pennsylvania Route 441, just east of the plant.

Shot of Three Mile Island from AMS/NEST helicopter. The pilot is on the left, the scientist is taking measurements on the right.

AMS/NEST helicopter on a monitoring mission.

Department of Energy Command Post, Capital City Airport. Left to right, L. Joe Deal, Herbert F. Hahn, and Wayne Adams.

Wall of command post showing organization chart, 1976 background survey, and monitoring station maps.

L. Joe Deal (1.). briefs Eric Beckjord and Andrew Pressesky of the Office of Energy Technology early in April.

A member of the Department of Energy's radiation assistance team analyzing samples at the command post.

Loading a fifty-five gallon drum containing a sample of reactor core coolant. The drum was shipped to the Bettis Laboratory for analysis.

President and Mrs. Carter tour the Unit 2 control room with Harold Denton (1.) on April 1.

L. Joe Deal, director of the Department of Energy's activities at the command post.

View of Unit 2. Cylinder-shaped containment building is in the background and the square auxiliary building in front left.

Placing a monitor with a local resident.

T. P. Stuart, senior scientist at EG&G, Inc., Nevada, preparing to count radiation levels collected by an air sample.

Brookhaven scientists logging in samples collected during the day.

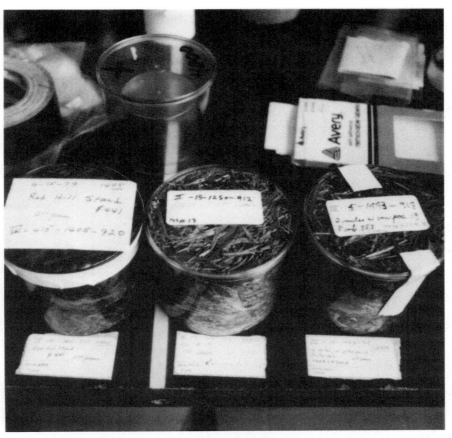

Examples of terrestrial samples collected on April 15, 1979.

Steven Gage of the Environmental Protection Agency during a tour of the command post.

Pennsylvania Department of Transportation hangar at Capital City Airport. The large containers on the trucks contain communications and technical equipment.

Briefing visitors at Capital City Airport. Left to right, Herbert F. Hahn, Commissioner Richard T. Kennedy of the Nuclear Regulatory Commission, and L. Joe Deal.

Five o'clock briefing. Gene Start of National Ocenographic and Atmospheric Administration is giving weather briefings.

Chapter V
The Battle of Three Mile Island

*There were two different levels. One was the
White House department head level which
had one perception of what was happening,
what was going on, and what should be going
on. Then, the scientific level of people knowing what they had to do, and the resources
and they were up there digging in the
trenches and doing a good job.*

John Villforth, Department
of Health Education and Welfare

*I don't know whether to cry my eyes out,
scream my head off, or wet my pants.*

Baby poster on the wall of
Three Mile Island Control Room

On Saturday, March 31, Three Mile Island became a confused battleground in which two armies, almost unaware of each other's existence, laid siege to the crippled reactor with its real or imagined dangers: radiation, explosion and meltdown. The first army consisted of the scientists and radiation experts from around the country who began arriving as
soon as the accident began on Wednesday, acting according to plan. The second army was a marvel of improvisation, an invasion of government officials
and reporters dispatched from around the country and from Washington.
Throughout the weekend these two armies, scientific and non-scientific,
deployed their troops and equipment almost independently of one another.
The scientific response was planned; the non-scientific response was not.

As on Friday, media amplification of public statements by government officials generated a sense of apocalyptic crisis. At the center of fear was the
hydrogen bubble, whose existence and unknown size and composition triggered rumors of meltdown, explosion and massive evacuation. These rumors
culminated at eight-thirty Satuday evening in the following Associated Press

wire story:

> Harrisburg, PA (AP). Federal officials said Saturday night that the gas bubble inside the crippled nuclear reactor at Three Mile Island is showing signs of becoming potentially explosive, complicating decisions on whether to mount risky operations to remove the gas.
>
> Officials said earlier that tens of thousands of people might have to be evacuated if engineers decided to try to remove the bubble, operations that could risk the meltdown of the reactor and the release of highly radioactive material into the atmosphere.
>
> But the NRC said Saturday night that it might be equally risky not to try the operation, because the bubble showed signs of gradually turning into a potentially explosive mixture that could wreck the already damaged reactor.[1]

On Friday, misinformation and poor communications created a situation of calm at the plant and fear in Washington. On the weekend that fear, fed by rumors such as those carried in the Associated Press story, was brought to Three Mile Island by the invasion of reporters, scientists and politicians who joined the experts already there. Fears of a meltdown or explosion were thus passed down to Department of Energy scientists at Capital City Airport, leading to a series of hurried contingency plans for evacuation from the area. The Three Mile Island accident somehow became magnified over the weekend because Washington came to Three Mile Island.

To an anxious public, the main event of the weekend was the President's hurried trip to the plant on Sunday afternoon. The visit marked the height of the political response and became the major media event at Three Mile Island. But at center, things were calmer; plant workers having beers, sandwiches or peanut butter ice cream at Kuppy's Diner in Middletown felt that the reactor was now stable and that it would soon be brought to a safe shutdown.

☆ ☆ ☆ ☆ ☆ ☆ ☆ ☆ ☆ ☆ ☆ ☆ ☆ ☆ ☆ ☆

The major problem at Three Mile Island was a large hydrogen bubble that had formed in the head of the reactor vessel. Small amounts of hydrogen are generally present in any pressurized water reactor in the steam space at the top of the pressurizer. Excess pressure can generally be "burped" or vented through a pressurizer by a spray nozzle—much like a bottle of soda pop. What was unusual and alarming at Three Mile Island was that hydrogen, produced when steam reacted with the overheated and uncovered zircalloy fuel rods in the reactor, had escaped from containment and into the primary coolant system itself. Government officals and utility operators were afraid that the hydrogen, if present with sufficient oxygen, might become flammable, or even explosive.[2]

Plant operators worked throughout the weekend on various schemes to reduce the size of the bubble (estimated at between eight hundred and a thousand cubic feet) and to cool the reactor core. A sizeable group of Nuclear Regulatory Commission officials and nuclear industry experts from around the country, the so-called Industry Advisory Group, or "Think Tank," brought to Pennsylvania by General Public Utilities' Herman Dieckamp, advised the men in the plant. Initially on Sunday radioactive gas was being vented through a temporary pipe. Later the standard operation of recombining hydrogen and oxygen into water was employed to eliminate the bubble. At the time of the President's visit on Sunday the Nuclear Regulatory Commission had enough accurate technical advice from around the country on the bubble's flammability and detonation potential to feel that a meltdown or explosion was virtually impossible. In Roger Mattson's words, "it's not going to go boom."[3] This growing technical confidence was not conveyed to the outside world, which continued to hear predictions of doom generated on the basis of Friday's confusion, miscalcuation and misinformation.

At eleven Saturday morning, Metropolitan Edison's Jack Herbein held his last press conference and announced that, "I generally think the crisis is over." This was promptly denied by Harold Denton, who said that the hydrogen bubble was larger than Metropolitan Edison's estimate and contained a dangerous and increasing level of oxygen. He implied a crisis situation that would last until the reactor reached a state of cold shutdown. Only late Saturday evening, at an eleven o'clock news briefing in Harrisburg held in response to media reports of impending disaster, did Denton maintain that an explosion was now impossible and evacuation unnecessary. "There is not a combustible mixture in the containment or in the reactor vessel," said Denton, "and there is no near-term danger at all."[4]

Washington's invasion of Three Mile Island thus placed a single federal spokesman in charge of all public relations for the utility and the government. Denton now began to brief the press and government officials on a regular basis. On Saturday morning the growing Nuclear Regulatory Commission encampment at Trailer City behind the Metropolitan Edison observation building across the river from the plant became the center for new information. At noon Denton held a press conference in Middletown at which he questioned Herbein's optimistic reports of a shrinking bubble and stated that no meaningful conclusions could be drawn as to its size. As Denton assembled a larger staff to deal with publicity, he began to schedule daily press briefings at the Middletown borough hall gymnasium.

Over the weekend the area around Three Mile Island took on the eerie quality of a battlefield deserted by the civilian population. Many residents had left the area, refugees from fear, and Middletown resembled a ghost town. The *Harrisburg Patriot* wrote of residents fleeing Goldsboro in droves and of "nuclear refugees" throughout Pennsylvania. Schools and businesses had closed, churches and homes stood empty, motel and restaurant employees were gone. The Rev. Clyde W. Roach of the Ebenezer A.M.E. Church said two-thirds of his congregation had left the area. "I cancelled this morning's service," he said, "but if I had preached it would have been along the lines that all things made by man are imperfect." On Sunday the Catholic diocese in the area granted general absolution to its members.[5]

Scientists monitoring off-site radiation that weekend found no cause for alarm. Environmental Protection Agency samplings of drinking water, food and milk turned up traces of iodine and cesium that were tens of thousands of times lower than Environmental Protection Agency protective action levels. Pennsylvania agriculture officials delivered samples to the Department of Environmental Resources showing equally low levels. "The readings have been negative," said Pennsylvania Secretary of Agriculture Penrose Hallowell; "so far there is zero radioiodine and only one test shows noble gases in tiny proportion, a barely measurable quantity." John Cope, vice president of Ashcombe Farm Dairies, also noted that there was no contamination. "We have monitoring devices with which we have been checking the area around our farms for radiation, and we've found no reading at all," he told reporters.[6]

Despite such reassuring statements, state and local officials were busy planning further evacuation to cope with potential disaster. The Pennsylvania Emergency Management Agency was planning for a massive evacuation by Tuesday, possibly out to twenty miles and involving some 600,000 people. Dauphin County officials, unable to reach Governor Thornburgh on Saturday, announced that they would begin their evacuation Sunday morning at nine; only a two a.m. telephone call by Lieutenant Governor Scranton succeeded in halting the planned move. Evacuation fears were fueled by the delivery of potassium iodide, an antidote to radioiodine, to Pennsylvania. But in the end, exodus preempted evacuation and officials did little more than belatedly publish escape routes in the *Harrisburg Patriot* on Monday.[7]

Many fears originated in Washington rather than in southern Pennsylvania. Throughout Saturday morning Department of Health, Education and Welfare Secretary Joseph Califano made inquiries regarding his Department's role in case of evacuation. At noon Califano sent a memorandum to the White House recommending a high-level meeting of cabinet officials and urging that the White House "make certain NRC closely consults with HEW and EPA public health experts." Despite a lack of information regarding conditions around the plant, Health, Education and Welfare officials in Washington favored evacuation out to ten or twenty miles and were busy arranging for the delivery of 240,000 potassium iodide vials to the Harrisburg area on Jack Watson's order. But Watson initially resisted the idea of a cabinet meeting on the grounds that it would be an unnecessary "Presidential media event."[8]

In Bethesda the Nuclear Regulatory Commission continued to be preoccupied with the bubble. Around noon on Saturday Joseph Hendrie called Denton, expressing concern that oxygen freed by radiolysis was building up in the reactor and urging Denton to warn Thornburgh of the danger. Shortly before three Hendrie held a news conference in Bethesda at which he noted that the reactor "continues to be in a stable configuration" but that "we do need to get the bubble out of the reactor vessel." He admitted that evacuation was "certainly a possibility" and answered reporters' questions about whether or not the bubble might explode and disperse radiation as far as Washington itself. In conference with other Commission officials Hendrie continued to talk of the "unknown" and the "edge of the precipice."[9]

Reactor safety expert Roger Mattson was equally disturbed. Around three-thirty Saturday he met with the Commissioners and warned that a buildup of oxygen could make the bubble flammable or explosive within a few days. "No plant has ever been in this condition," said Mattson; "no plant has ever been tested in this condition." Only Saturday evening did a report by James Taylor of Babcock & Wilcox reach the Nuclear Regulatory Commission. The report contained an engineer's earlier calculation that no excess oxygen was being generated, and that the danger was therefore much less acute than believed, but the report apparently never reached Mattson. Instead, what came out on the AP wire Saturday evening was reporter Stan Benjamin's interview with Nuclear Regulatory Commission officials stressing a potential explosion.[10]

The Three Mile Island accident became more explosive politically that weekend. Anti-nuclear rallies were held across the country. Consumer advocate Ralph Nader called for the immediate evacuation of all who lived within thirty miles of the plant. Senator Gary Hart, a Democrat from Colorado, declared on "Face the Nation" that nuclear power now had a "credibility gap." Governor Jerry Brown of California wired the Nuclear Regulatory Commission asking them to close down temporarily the Rancho Seco power plant near Sacramento. Dr. Norman C. Rasmussen of the Massachusetts Institute of Technology, author of an important report on the probability of nuclear power plant accidents, admitted that "we may have to change our projections somewhat" regarding future accidents. There were reports of "suicide volunteers" offering to enter the containment to shut down the reactor, of plans to disperse radioactive gas in balloons, and of evacuation for horses, cows and goldfish as well as people. "Stop the Merchants of Atomic Death," pleaded a banner outside President Carter's First Baptist Church in Washington.[11]

Mounting political pressure also forced the White House meeting which Califano had been urging. Chaired by Jack Watson, the meeting lasted until Saturday evening and included four officials from the Department of Health, Education and Welfare. No Deparment of Energy representatives were invited. None of the Health, Education and Welfare officials except John Villforth were aware of the Interagency Radiological Assistance Plan or of the Department of Energy's role at Three Mile Island. There was some discussion of the need to let state and local officials lead the way, and minimize public anxiety by avoiding statements to the news media. But the meeting ended by assigning to the Nuclear Regulatory Commission the major role for coordinating environmental monitoring data—the very role which the Department of Energy, at the request of Pennsylvania officials, was already playing. The White House remained unaware of the Department of Energy commitment, yet the Department continued to serve as the lead agency in off-site monitoring, completely unaffected by the high level decision.[12]

☆ ☆ ☆ ☆ ☆ ☆ ☆ ☆ ☆ ☆ ☆ ☆ ☆ ☆ ☆ ☆ ☆

The idea of a Presidential visit to Three Mile Island originated early Saturday evening. Presidential domestic affairs adviser Stuart Eizenstadt suggested the idea in a telephone conversation with President Carter who was aboard Air Force One en route to Milwaukee to give a speech. Eizenstadt then cleared the plan with Harold Denton. That evening the President announced his intention. Admitting that a "very serious problem" still existed and that the crisis was not over, even if "improving," Carter announced that he planned to visit Three Mile Island "in the near future."[13]

The near future was nearer than Jessica Tuchman Matthews, the White House nuclear adviser, knew. At two o'clock Sunday morning she was awakened at home by a telephone call and asked to report to her office and prepare a briefing memorandum for the President. Upon arrival at the White House she called Harold Denton in Pennsylvania, and then spoke with Zbigniew Brzezinski and Vice President Mondale. With Jack Watson, Matthews then wrote a memorandum urging the President to say nothing during his visit about the technical problem at the reactor and not to convey the impression that the danger was over.[14]

By Sunday morning Vic Stello of the Nuclear Regulatory Commission had concluded that no more oxygen was being added to the hydrogen bubble. Unfortunately, Roger Mattson and Joseph Hendrie, who had already left Bethesda for Three Mile Island to help Denton brief the President, carried obsolete data indicating that the bubble was now flammable and that more oxygen levels were rising. As Mattson recalled it, Stello told him he was "crazy" and that the Commission had made an error in calculating the bubble's explosive potential. "Stello says we're nuts and poor Harold is there; he's got to meet with the President in five minutes and tell it like it is . . . And here he is. His two experts are not together. One comes armed to the teeth with all these national laboratories and naval reactors people and highfalutin' Ph.D.s around the country, saying this is what it is and this is his best summary. And his other, [the] operating reactors division director, is saying 'I don't believe it. I think it's wrong.' "[15]

Against this background of confusion, President and Mrs. Carter got off a U.S. Marine helicopter at one o'clock Sunday afternoon at an Air National Guard airport near Three Mile Island.

At an airport briefing, Denton gave both Mattson's and Stello's views on the reactor. The Carters, Denton and Governor Thornburgh then boarded a school bus and headed for the plant. The Presidential party donned yellow plastic booties and made their way through the crippled plant to the Unit 2

Control Room. There Jim Floyd and Harold Denton explained the current status of the reactor.

After touring the plant, the President returned to Middletown for a press conference at the borough hall gymnasium. Hand-shaking his way through a cheering, if small, crowd on West Emmaus Street, the President confronted several hundred reporters in a forest of microphones, cameras, cables and glaring lights. Together with Denton and Thornburgh, Carter urged calm and promised that "everything can and will be done in coping with the problem and those that might occur in the future."[16]

The Carter visit helped calm a difficult situation. Most Middletown residents still remaining in the area found the President's visit reassuring. At lease one did not. A forty-five-year-old iron worker, Carl Lankert, reportedly remarked: "What has he got to do with all this? It's just good politics."[17]

The Presidential visit was in fact the result of mounting pressure in Washington, especially from the Department of Health, Education and Welfare, to demonstrate high-level political involvement at Three Mile Island and a visible concern for public health and safety. Until the time of the visit, however, Nuclear Regulatory Commission officials still thought that the bubble was flammable and potentially explosive, that a meltdown was theoretically possible, and that evacuation was therefore a possibility; officials in Harrisburg were discussing evacuation contingencies even as the President toured the plant. We know now that the dangers of the bubble exploding were based on rumor or false calculation. The President's Commission concluded that "throughout Sunday the NRC made no announcement that it had erred in its calculations or that no threat of an explosion existed."[18]

The Presidential visit characterized a Washington response to the Three Mile Island accident that was sometimes *ad hoc* and political, rather than preplanned and scientific. A chain of misinformation, confusion and poor communications stretched from the distant offices of the federal government to Harrisburg and the plant. But there was also a parallel chain of command stretching from distant laboratories in California, Idaho and Nevada through Germantown to Three Mile Island involving scientists from around the country. Contrary to Mattson's comments about "highfalutin' Ph.D.s," laboratory scientists were accurately assessing the plant's difficulties. Unfortunately, this information was not properly used. This second chain of command remained almost invisible to the press and the politicians. Riding into Middletown like

the U.S. cavalry, they failed to see the weary vanguard that had been en-
camped for days on the battlefield.

☆ ☆ ☆ ☆ ☆ ☆ ☆ ☆ ☆ ☆ ☆ ☆ ☆ ☆ ☆

By Saturday the Department of Energy had quietly established itself, at the
request of the Nuclear Regulatory Commission and Pennsylvania officials, as
the main federal agency involved in monitoring, analyzing and coordinating
the increasing amount of data on off-site radiation from the plant. Over the
weekend the Department became involved not only in radiation monitoring,
but at the Nuclear Regulatory Commission's request, in the ferrying of lead
bricks from Pittsburgh and Long Island to shield the recombiner machines
employed to reduce the hydrogen bubble and in the transport and analysis of
samples from the primary coolant system. The Department also made plans
to evacuate all Nuclear Regulatory Commission and Department of Energy
personnel from the area if necessary and analyzed detonation potential and
dispersal possibilities of the hydrogen bubble. The basis of this expanded role
was not planned, but resulted from requests from other agencies for
assistance.

Pilots and scientists continued to fly plume flights near the plant on a regular schedule. From six to seven-fifteen a.m. Saturday morning Tom Maguire
flew out thirty miles from the plant going along the plume but did not encounter readings of more than one millirem. Three more flights Saturday
revealed no major changes, no iodine and no readings above three millirems.
In general, low off-site radiation readings persisted, in marked contrast to
rumored dangers of explosion and meltdown emanating from public officials
and the media.[19]

Plume flights were conducted on a three-hour basis. Helicopters had to
compete for air space around the plant with Metropolitan Edison and the
media. At times as many as seven helicopters were in the air simultaneously.
Pilots drew straws to see who would fly at night when conditions were dangerous; those who won, or lost, were rewarded after their flights with beer and
sandwiches, courtesy of Tony Sauder at the Red Baron restaurant at Capital
City Airport. Even during routine flights the adrenalin flowed freely, and the
men grew weary.[20]

In addition to plume flights, there were other sources of off-site radiation
monitoring data. On Saturday, some thirty people in different Department of
Energy teams were undertaking terrestrial monitoring in the area. This was

usually done by car, with drivers and escorts provided by the Pennsylvania State Police with the cooperation of the Bureau of Radiation Protection. Most terrestrial monitoring utilized scientists from Brookhaven and Bettis Laboratories, riding in assorted station wagons, vans and four-wheel-drive vehicles, using radiation detectors and bringing samples back to Capital City Airport for analysis.[21]

Analysis of local weather data by ARAC also began on a regular basis Saturday. Since Thursday Marv Dickerson of the Livermore Laboratory had been at Capital City Airport making calculations based on wind, weather and plume readings and relaying his information to both the Department of Energy and the Nuclear Regulatory Commission. On Friday ARAC was installed at the airport, so that hourly predictions of weather, plume direction and dosage could be obtained by telephone directly from the ARAC computer at Livermore. By Saturday, twenty-five people at the laboratory were involved, under the direction of Christine Sherman, in making computer calculations for Three Mile Island.

Radiation monitoring and data analysis were by now routine. But on Saturday the Department of Energy was called upon to play a broader role in responding to the accident than emergency planning had anticipated.

☆ ☆ ☆ ☆ ☆ ☆ ☆ ☆ ☆ ☆ ☆ ☆ ☆ ☆ ☆ ☆ ☆

Don Ross did not sleep much on Friday night. At three-thirty Saturday morning Bill Ward of the Nuclear Regulatory Commission called Ross at the Emergency Operations Center and informed him of a new problem. Metropolitan Edison had requested the Commission to obtain as many lead bricks as possible to shield recombiners and other equipment to be used in reducing the size of the hydrogen bubble. They were asking for two thousand lead bricks in a matter of hours. "Fast is the word," said Ross.[22]

Ross promptly called the Command Post, the Bettis Laboratory in Pittsburgh and the Joint Nuclear Accident Coordination Center in Albuquerque. By 5:05 he had learned that Bettis had a thousand bricks stored two miles from the airport in Pittsburgh. He informed the Nuclear Regulatory Commission, then called O. J. Woodruff at Bettis and asked him to transfer the bricks to Allegheny Airport for shipment to Three Mile Island on an Air Force C-130.[23]

The request for lead bricks was passed on to Brookhaven that night. By seven Saturday morning two flatbed trucks, each with twenty thousand pounds of lead bricks, began rolling toward Three Mile Island. By Saturday afternoon, another 89,200 pounds of lead bricks were ready at Westhampton Air Force Base for shipment to Three Mile Island by the Military Air Transport Command. At 7:45 Sunday morning the Nuclear Regulatory Commission reported that enough bricks had been delivered (400 tons) and they did not need any more. Work began on the shielding for the hydrogen recombiner.[24]

Don Ross received another jolt from the Nuclear Regulatory Commission Saturday morning. Shortly before six Leo Higginbotham of the Commission had asked Ross what contingency plans the Department of Energy had in the event of a core meltdown at Three Mile Island. Within half an hour Bernie Weiss also called to inquire about Department plans for an escalated emergency. Ross told Higgenbotham that he could call him back by eight, but Weiss did not want to wait. Given this new sense of urgency, Ross promptly called Roy Lounsbury.[25]

Colonel Roy Lounsbury was the man who would bring military contingency planning to the battle of Three Mile Island on Saturday. Since October 1976, Lounsbury had worked for the Department as manager of the Emergency Operations Center and coordinator of NEST operations. The Center and its Emergency Action Coordinating Team operated under the Office of Military Applications. In the event of any emergency, the director of the Team would be from that Office. But except for the 1978 Operation Morning Light search for the downed Soviet satellite in Canada, the system had never been tested, nor had there been much coordination with the Nuclear Regulatory Commission. Morning Light had provided valuable experience in a quasi-military radiological operation, and there was coordination at Las Vegas between NEST and Environmental Protection Agency personnel.[26]

Lounsbury and his staff had learned lessons from Operation Morning Light that would be crucial at Three Mile Island: there must be a single Department field representative on site in any emergency; whenever an operation was deployed, prepare a staff for at least twenty-four hours; bring your own communications system, including AT&T long-line telephones; bring more than you need. In addition, isolation from the press was essential to efficiency. These lessons would be invaluable at Three Mile Island.

Lounsbury had first heard of the accident at Three Mile Island late Wednesday morning at the Pentagon. At first he had felt that it was, as he put it, "nothing for me to get excited about." Later that afternoon he went out to Germantown, visited the Emergency Operations Center, talked to Jack McCool, and recommended that a senior Department person be put in charge at Three Mile Island. By this time Lounsbury encountered an atmosphere of "calm confusion" at the Emergency Operations Center and realized he was dealing with a nuclear accident of serious potential. Since it was not a weapons-related accident, General Joseph K. Bratton designated McCool of Environmental Safety as the man in charge of the Center. But until Saturday Lounsbury saw no need to go to Three Mile Island himself.

What changed the situation was a telephone call from a Nuclear Regulatory Commission official to Joe Deal's motel room at six Saturday morning. Did the Department have a place to move its operations in case of a serious radiation release? asked the caller. No, grumbled a sleepy Deal. The command post might not be far enough away, and contingency plans for evacuation of personnel might be in order. "By the way," said the caller, "while you're at it, if you can find a place for our people to go, they'll go with you." Deal promptly dressed, drove out to the command post and telephoned General Bratton, who decided it was time to send Lounsbury and Walt Plummer of the Office of Military Application to Three Mile Island.[27]

Half an hour after Deal had been rudely awakened from a sound sleep, Don Ross at the Emergency Operations Center reported the heightened Nuclear Regulatory Commission concern to Lounsbury. Asked if there were any additional plans for an escalated emergency, Lounsbury replied that "for what they're calling the meltdown, we would just have to find—we would provide whatever is needed in the way of RAP [radiological assistance]; that's all we can do. We don't have anything else. We're not going to be bringing in five hundred people," said Lounsbury, "because there is nothing five hundred people can do."[28]

Around seven Ross put out a situation report on the accident, which did not reflect the concern for contingency planning: the reactor was stable and cooling, but "to avert a potential H_2 explosion hazard, the Nuclear Regulatory Commission would like to start the hydrogen recombiner." Off-site sampling still showed no iodine in milk, no detectable radiation in downstream filtration plants, and very low radiation levels.[29]

By eight o'clock Saturday morning the command post and the Nuclear Regulatory Commission were requesting Emergency Operations Center aid in planning for a new emergency. Joe Deal told Jack McCool that he wanted Lounsbury up at Three Mile Island as soon as possible. Bernie Weiss of the Nuclear Regulatory Commission told McCool that "people are getting a little pessimistic in the sense that we ought to do some more contingency planning in case something much worse happens." The problem was not to obtain more people, but to design an evacuation plan for government personnel. The Nuclear Regulatory Commission's alarm was spreading. "They're thinking about evacuation in the event of a large release," McCool told Deal; "They're thinking about the meltdown and they want to be prepared to move people out in a hurry if possible."[30]

If the Nuclear Regulatory Commisssion in Washington wanted to move people out, more scientists were moving in. The command post was bustling. Four separate teams from Bettis, Argonne, Brookhaven and Oak Ridge were out doing terrestrial off-site monitoring under the general direction of Bob Friess. Bettis and Brookhaven teams also established portable laboratories to analyze the samples at Capital City Airport. Oak Ridge and Argonne personnel were assigned to the Nuclear Regulatory Commission for field monitoring. In addition, Joe Deal had divided operations into three general areas: terrestrial radiological monitoring under Friess, AMS/NEST aerial monitoring under Jack Doyle and ARAC operations under Marv Dickerson. Capital City Airport remained the center of Department of Energy operations. Deal later testified that the airport was "easy for everybody to get to and it was private in the sense that we [had] made arrangements with the Capital City Airport Chief of Police [Donald Grow] to provide us guard service, and so we weren't interrupted by people we didn't want around the place."[31]

Establishing the command post at Capital City Airport turned out to be a wise and prudent decision. Supplies and personnel could be brought in directly by air to the place where plume flights and other monitoring operations originated; contact with headquarters and the national laboratories was assured by extensive dedicated telephone lines established on Wednesday. In contrast Joe Deal observed that, in Harrisburg, Pennsylvania officials were "absolutely paralyzed" by almost constant requests to brief the Governor and the press on the operations. The Nuclear Regulatory Commission also had "press everywhere; guys sticking microphones under them and cameras working around." In addition, Deal noted that being on the scene, rather than in Washington, was crucial. "It can't be planned from down there," he told

Jack McCool, "and that's why we need Lounsbury's help."[32]

By mid-morning Saturday Lounsbury and Walt Plummer were in an Air Force helicopter flying north over the rolling Maryland countryside on their way to Three Mile Island. Their mission was to establish an alternate location for the command post in case evacuation became necessary. The most likely spot was the U.S. Army Post at Carlisle, Pennsylvania, twenty miles west of Harrisburg, and Lounsbury had already arranged its availability by telephone. Upon arrival at Capital City Airport, Lounsbury called Scranton's office and asked if any state officials wished to participate in arranging for relocation; they did not. After briefings by Deal, Hahn and Doyle, Lounsbury and Plummer enjoyed lunch at the Red Baron, and headed off by helicopter for Carlisle.[33]

By late Saturday afternoon, Lounsbury and Plummer had inspected the Carlisle Barracks, and General Bratton had reserved half of Building 684 in case of evacuation. In addition, Lounsbury had decided upon the code name "Jim Thorpe" for such an operation, named after the fabled Indian All-American football player from Carlisle College. If evacuation were to be ordered, the name "Jim Thorpe" would be passed along to all Department personnel at Three Mile Island. This plan was first unveiled by Lounsbury at a "worst case contingency planning meeting" held at Pennsylvania Emergency Management Agency headquarters in Harrisburg, complete with a large wall photo of a mushroom-shaped cloud.[34]

The relocation plan called for all Department personnel and scientists to move to Building 684 at Carlisle Barracks. National laboratory scientists would move first, command post personnel second, and aerial monitoring teams last. Each group was expected to provide its own cars or other means of transportation. Flight operations would continue as long as possible, directed by the senior EG&G scientist, Lounsbury said. Carlisle itself would feature U.S. Army cuisine and "humble but adequate sleeping facilities." In addition, a Department memo emphasized that in an emergency the Department could bring in additional doctors, health physicists and scientists with experience in serious nuclear accidents.[35]

When Lounsbury repeated his briefing at the command post later, he created a stir. Lounsbury was a military man used to contingency planning. In his mind, he was giving a kind of "morale booster" speech that said, in effect: Don't worry because if the worst happens, we are prepared for it. He outlined the threat of a hydrogen explosion. He described the evacuation plan. He urged all Department personnel and scientists to be ready to move at any time and to have their suitcases packed. According to Lounsbury, there was little overt reaction to his briefing, only a mood of prudent preparation.[36]

Lounsbury's briefing had a more serious impact than he knew. Joe Deal had been alarmed by Nuclear Regulatory Commission talk of a meltdown and had had Marv Dickerson prepare a hypothetical plume pattern and map overlays for display at the briefing.[37] This visual worst-case display upset Pennsylvania officials, because the radiation appeared to spread well beyond the twenty mile distance for which they had any evacuation plans. Others interpreted Lounsbury's plan as "an exercise." Air Force pilots Creel and Bauland were military men and equally unconcerned; to them it was just "a standby plan."[38] But Tom McGuire recalled thinking about "the real possibility of some kind of danger." If an Army colonel was there, "these guys are playing for real," he thought. Bob Casey could not remember Lounsbury's last name, but remembered a "very strong concern" about possible meltdown. Bob Friess had trouble sleeping afterwards and recalled that "a lot of us were a little uneasy;" Saturday, he felt, was definitely the "crisis day."[39]

Bob Shipman, who had been at Three Mile Island since Wednesday, was also alarmed by Lounsbury's speech. The radiation levels he had recorded seemed far from dangerous, but here was a military man talking about meltdown, evacuation, and potassium iodide capsules. "That was scary," said Shipman. "I had this sudden vision of all these DOE people running like rabbits out through the hills trying to get away from this green cloud."[40] Lounsbury's very presence now lent credence to rumors and press reports of explosion and meltdown.

Washington's fears had come to Three Mile Island.

☆ ☆ ☆ ☆ ☆ ☆ ☆ ☆ ☆ ☆ ☆ ☆ ☆ ☆ ☆ ☆

At one-thirty Saturday afternoon Jack McCool called Department of Energy Secretary James Schlesinger from the Emergency Operations Center and gave him an update on the situation. McCool reported the latest information concerning the bubble in the top of the pressure vessel, noting that "the radioactivity in the primary coolant is quite high. The radioactivity off site is quite low." Radiation readings five miles from the plant were about one millirem, showing xenon and no iodine. Schlesinger was not greatly concerned and asked few questions.[41] Unlike Califano, his direct involvement in the

Three Mile Island response was minimal. Intervention from Washington did not seem necessary.

More significant that afternoon was the Department of Energy contribution to the hydrogen bubble problem. This contribution took two forms: delivery of a sample of highly radioactive gas from the bubble to the Bettis Laboratory in Pittsburgh, and an analysis of various strategies for venting the bubble, carried out at the Department's Idaho National Engineering Laboratory.

At 7:45 a.m. Saturday Joe Lenhart of Oak Ridge called the Emergency Operations Center and reported that the plant had taken a sample of gas from the bubble and needed an analysis of its hydrogen content. Lenhart recommended that the analysis be done at Bettis Laboratory in nearby Pittsburgh, rather than at Oak Ridge. He asked Don Ross to call the plant to get the sample released to the Department of Energy, something Ross felt could be easily done since "apparently Schlesinger called the plant manager and told him he expected cooperation so we should have no trouble getting cooperation." Ross also call Joe Deal to set up the lab analysis with Bettis and arrange transportation. By nine Jack McCool had scheduled the Bettis analysis and informed Lenhart. He had also instructed Ed Patterson to pick up the sample at the plant and told Jim Floyd that Patterson was coming.[42]

At 11:25 a.m. Bernie Weiss called McCool and said he had a second sample drawn from the reactor's primary coolant loop to be analyzed. Again, McCool decided to do the analysis at Bettis rather than Brookhaven or Oak Ridge. All of this became academic since the Nuclear Regulatory Commission did not provide the sample until later.[43]

At 10:49 Saturday night Bettis telephoned the results of the gas sample to the Emergency Operations Center; the gas was showing mainly fission products xenon and iodine.[44] More significantly, early Sunday morning the lab called in the results of the mass spectrometry analysis for the second sample of primary coolant and reported very little uranium, signifying essentially no fuel melting in the reactor. This was important evidence that the accident was less serious than had been feared.[45]

Two thousand miles away from Three Mile Island the Idaho National Engineering Laboratory sprawls across 893 square miles of Idaho sagebrush. The laboratory sits upon the Snake River plain, a massive volcanic rock formation whose run-off water is home for ninety percent of the rainbow trout sold in American stores and restaurants. Antelope graze within a few hundred yards of the cement-block buildings. A major function of the laboratory is to reclaim nuclear materials from spent fuel rods used in reactors, especially in Navy submarines. The laboratory also possesses computerized semiscale reactor mock-ups to simulate accidents. In December 1978 it completed a series of successful tests on the response time of emergency core cooling systems under various accident conditions.[46]

Work on the bubble program was well underway there by Saturday. All Friday night, technicians had modified the blue-and-yellow semiscale reactor to conform to the conditions at Three Mile Island-2 and injected nitrogen gas to simulate the eight-hundred-cubic-foot bubble. "The entire crew volunteered to work on the experiments," recalled Larry Ybarrondo; "We had more people than we actually needed." At about seven o'clock Saturday morning the test began. By noon the test was completed and engineers were starting to evaluate the data. That evening Ybarrondo and Willis Bixby, at the request of the utility's Herman Dieckamp, called Vic Stello of the Nuclear Regulatory Commission at the plant site and agreed that Ybarrondo would come to Three Mile Island as soon as possible. At ten p.m. (Mountain Standard Time) the test results—240 separate printouts from the computer called Puff, the Magic Dragon—were sent to the Nuclear Regulatory Commission by telefax. Thus by midnight (Eastern Standard Time) the Department had provided the Nuclear Regulatory Commission with its requested bubble analysis: the semiscale model had successfully forced a bubble of gas into the outlet pipe of the coolant system and into containment, lowering pressure without uncovering the fuel rods.[47]

Throughout Saturday Idaho scientists were also busy devising ways to collapse the bubble and vent the hydrogen. "Imagine a pressure cooker with boiling water that is turning into steam," said Willis Bixby to an eager reporter. "If you leave the lid off, eventually the water will boil dry. But if you close the lid the pressure will increase. In order to lower the pressure without boiling it dry, you have to reduce the heat." Four options were considered: 1) to raise the pressure inside the reactor to collapse the bubble by dissolving it into the water of the primary coolant; 2) to lower the pressure, further exposing the core, in order to siphon out the bubble along with the primary coolant water; 3) to drop the water level and then reflood the reactor with fresh water, sinking the bubble; 4) to restart the reactor, creating heat sufficient to turn the water into steam and bursting the bubble. The fourth option was almost immediately rejected on grounds that restarting a reactor with possibly damaged fuel rods was both dangerous and impractical. Ultimately, the solution involved none of these options but the standard use of recombiners to turn the hydrogen and oxygen back into water.[48]

All day Sunday the Department continued to aid federal and state efforts at Three Mile Island. Helicopter plume flights continued every three hours around the clock, despite rain and dense fog. Two scientists from the Department's Environmental Measurements Laboratory in New York arrived by van and began monitoring around the plant area. AMS/NEST communications pods arrived from Las Vegas: airline cargo containers holding telephones, radios, handi-talkies, TV receivers, a telecopier and other communications equipment. Dick Beers arrived from Las Vegas to supervise some thirty EG&G personnel involved in AMS flights. And again, the five o'clock Department briefing at Capital City Airport provided an opportunity for agencies to exchange views and data. Yet the President's Sunday visit did not include the major off-site facility at Three Mile Island: the Department of Energy command post. The Department's profile remained low.[49]

☆ ☆ ☆ ☆ ☆ ☆ ☆ ☆ ☆ ☆ ☆ ☆ ☆ ☆ ☆ ☆

Only on Monday did most scientists and officials realize that the dangers of the hydrogen bubble had been overestimated, and that it was now greatly reduced in size. Relieved Nuclear Regulatory Commission officials in Bethesda said privately that they were "not concerned about the bubble and any explosion problem with it" and even wondered "whether there ever has been any substantial bubble." At an eleven fifteen a.m. news conference in Middletown Harold Denton reported a "dramatic decrease in the bubble size" (from 850 to 50 cubic feet), a drop in off-site radiation levels and "reason for optimism." The daily Nuclear Regulatory Commission bulletin also admitted that "further analyses and consultations with experts has led to the development of a strong consensus that the net oxygen rate inside the noncondensible bubble in the reactor is much less than originally conservatively estimated." This was the closest the Commission came to admitting that it had been wrong.[50]

Yet public concerns created over the weekend were not so quickly dissipated. Schools in the Harrisburg region remained closed until Wednesday, evacuation routes were published in the *Harrisburg Patriot* and only seventy per cent of Harrisburg's state employees showed up for work (6,500 absentees, as opposed to the usual 2,500). Many depositors lined up to withdraw money from local banks. "Gas Bubble Still a Hazard," read a *Washington Post* headline; "the battle to avert a major disaster" was still raging, said the *Wall Street Journal,* and "officials indicated that they need to come up with a solution within the next several days or conditions within the reactor might reach an explosive state." By noon Monday, however, reporters knew that the bubble was nearly gone. Monday's stories reflected the weekend's fears and confusion, not the latest reports from the crippled reactor.[51]

After all the activities of the weekend, the solution to the bubble problem was routine: spray nozzles in the pressurizer removed hydrogen-laden coolant through a nozzle in the top of the pressurizer, and Monday afternoon the recombiners began to convert excess hydrogen and oxygen back into water. Whatever the nature of the bubble, by Monday it was being eliminated.

☆ ☆ ☆ ☆ ☆ ☆ ☆ ☆ ☆ ☆ ☆ ☆ ☆ ☆ ☆ ☆

The Department of Energy remained active on Monday. Terrestrial monitoring continued, but plume flights were cancelled due to more bad weather and poor visibility. A C-141 was put on standby to carry a fifty-five gallon drum with a primary coolant sample to the Bettis Laboratory for analysis. The daily five o'clock briefings continued at Capital City Airport, complete with aerial photographs, topographical maps and radiation iospleth maps. Evacuation plans remained in effect, however, and the Nuclear Regulatory Commission now requested Department assistance in cleaning up contaminated buildings and equipment at Three Mile Island. By now there were over one hundred department personnel operating out of Capital City Airport.[52]

Problems persisted. Getting accurate up-to-date information from inside the plant and from the Nuclear Regulatory Commission remained difficult. According to Pennsylvania Secretary of Health Gordon McCleod, who visited the command post Monday, reports of iodine in milk samples had caused a number of farmers in the area to evacuate, leaving their dairy herds behind. Dr. Walter H. Weyzen of the Department's Office of Health Studies and Environmental Research also reported a "lack of liaison between the Department and state health authorities, and the local lack of expertise on either side to utilize monitoring information in medical decision making."[53]

☆ ☆ ☆ ☆ ☆ ☆ ☆ ☆ ☆ ☆ ☆ ☆ ☆ ☆ ☆

The major problem over the weekend at Three Mile Island was communication between scientists and politicians, the site and Washington, the plant and off-site experts. More particularly, it would appear that commands and statements were transmitted in profusion from Washington to scientists at Three Mile Island, but scientific data gathered through monitoring and laboratory analysis were not effectively understood in Washington. The army of scientists had plenty of guidance from its officers, the politicians, but the officers operated without the intelligence gathered by their army.

By early Sunday morning, Department laboratory analysis should have convinced the Nuclear Regulatory Commission that the hydrogen bubble was of little danger and could be vented. In the case of Vic Stello, it apparently did. But that analysis was going directly from Idaho to the Nuclear Regulatory Commission, not to the Department itself; Jack McCool reported Monday morning that he had learned of the Idaho simulation tests on the bubble only from watching television.[54] Because of poor communications between the plant itself and the command post, information often came from the national media, rather than directly from industry or government.

By Monday the quick and accurate communication of technical data to a fearful public and a distant federal government was better. Harold Denton was an effective spokesman to the press and the public, a valuable consequence of Washington's invasion of Three Mile Island. Public concern lessened and refugee residents began to return home. Scientists from industry and government were at the plant advising on procedures for a cold shutdown. The battle of Three Mile Island continued on new fronts.

Chapter VI
The Changing of the Guard

You Americans don't solve problems, you overwhelm them.

Canadian official
at Operation Morning Light, 1978

As the weekend ended, it became clear that the immediate crisis had also ended but that a long campaign to bring the reactor under control was about to begin. Temporary expedients became permanent procedures. Fresh recruits replaced grizzled veterans. Those who had thought the crisis had been checked on Wednesday or Thursday now realized they were in for a long stay. Slowly residents returned to the area and their daily routines, but for days schools remained closed and motel rooms filled with scientific and official visitors from around the country and the world. The "Harrisburg Syndrome" continued to make headlines even as the less glamorous tasks of cleanup began.

At the Middletown gymnasium, Harold Denton and a staff of tired and overworked Nuclear Regulatory Commission and Department of Energy public relations personnel held twice-daily press briefings amid a forest of bright lights, cables snaking across the floor, and hungry reporters trying to meet deadlines. At times the world's entire press corps seemed to be jammed into the old building at one time, asking endless questions about the reactor and trying to understand the jargon of nuclear reactor technology. By Tuesday a number of Department of Energy public relations people, including James Cannon, Gail Bradshaw and Robert Newlin of headquarters and James Alexander of Oak Ridge, assisted the Commission's press staff. Conditions were not ideal — the xerox machine was in the kitchen and the daily bulletins were handed out from the pantry after being collated on chairs set up along the gymnasium wall. By the end of the first week after the accident, a regular system had developed for answering, or trying to answer, endless questions about nuclear technology and terminology.

Trailer City had become a semi-permanent encampment behind the observation building. The city possessed a helicopter landing pad, a large mess tent and dozens of light-green mobile homes, which spilled over the parking lot and into the surrounding fields. From here the Nuclear Regulatory Commission provided information and periodic bulletins for officials, reporters and scientists. In addition, a mimeographed newspaper, "Trailer City News," kept the camp informed of local events. Yet the information available at the gymnasium press center and Trailer City included little publicity on the extensive and continuing Department of Energy effort.

A week after the accident had begun, hundreds of scientists and technicians from the Department of Energy and its contracting laboratories continued to invade the Three Mile Island area. Fresh troops arrived every few days to relieve those who had spent long hours in the field. Soil, vegetation, water, air and milk samples continued to be reassuring: no unusual radioactivity was found, and the scientists agreed that public safety had not been threatened. Aerial monitoring teams were still spotting xenon, but at very low levels. In fact, only the extreme sensitivity of their instruments enabled them to pick up the low radioactivity they did register. Certainly, they felt, there was no cause for alarm. Accustomed to responding to a clear and present danger, the radiological assistance teams chafed under what they now perceived to be a purely political response. But each day, like good soldiers, the men set out on their monitoring missions, patrolled the plume, brought back captive samples for analysis and reported their findings to the Nuclear Regulatory Commission or Pennsylvania officials. They could report that the enemy, dangerous radioactivity, remained under siege within the looming fortress on Three Mile Island. And for them the urgency of battle had begun to disappear.

Like any other military operation, the government's response to Three Mile Island needed a code name. Roy Lounsbury knew that the Pentagon reserved certain letter groups for non-military events. He phoned to check. He could choose any combination of two words beginning with the letters "I" or "P." For a while Lounsbury toyed with possibilities, jotting several down on a sheet of paper. With his list of options before him he called the Nuclear Regulatory Commission. The major responsibility at Three Mile Island was theirs, he reasoned, and someone from the Commission should choose the operational code name. He read off his list. Later that day someone from the Commission called to tell Lounsbury what name had been selected. Rejecting all of Lounsbury's suggestions, the Commission had settled on Operation Ivory Purpose. The Air Force helicopter pilots had also christened the event. As the crisis wound down after April 1, they began to call it Operation Faded Glory.[1]

☆ ☆ ☆ ☆ ☆ ☆ ☆ ☆ ☆ ☆ ☆ ☆ ☆ ☆ ☆ ☆

The first few days of April were damp and dismal. Angry gray clouds scudded low across the ridges north of Harrisburg. Heavy fog reduced visibility and hampered the aerial monitoring flights. Ground teams, however, continued patrolling the perimeter of the plume, collecting samples. Other crews established radiation dosimetry stations within a twelve-mile radius of the plant to obtain a cumulative dose assessment, as requested by the Nuclear Regulatory Commission. The command post possessed the most reliable telephone communications in the area and often coordinated the Nuclear Regulatory Commission and Metropolitan Edison requests for additional support.[2]

Away from the command post, additional Department of Energy scientists were exploring and analyzing the reactor's problems and attempting to find possible solutions to them. Larry Ybarrondo had arrived to join the "think tank" of experts advising the utility. And on Three Mile Island stood Herman, the sturdiest soldier in the Department of Energy's army, known to the scientists at the Oak Ridge National Laboratory as a "mobile manipulator." Jargon aside, Herman would travel where others could not go, a sophisticated robot who marched unhesitatingly into areas of deadly radiation. Metropolitan Edison had asked that Herman and his human staff be trucked in to take radioactive samples from the damaged reactor's core coolant. In dealing with the most serious commercial power reactor accident in the industry's history, the Department of Energy was holding back none of its vast resources in the effort to protect the public.[3]

The Department of Energy's effort was far more organized by April 2 than it had been even a few days earlier. Communications equipment so badly needed when the existing telephone circuits had become overloaded finally arrived from Las Vegas despite an airline strike which had delayed the shipment. After the equipment arrived, life at Capital City Airport became much simpler.[4]

With the additional Nevada equipment, the command post was able to serve as the focus for all governmental monitoring activity at Three Mile Island. At Joe Deal's suggestion, the scientists and technicians from the Department of Energy, the national laboratories, AMS/NEST, the Environmental Protection Agency, the Food and Drug Administration, the Nuclear Regulatory Commission and the Pennsylvania Bureau of Radiation Protection continued to meet each day in the airport hangar behind the command post to brief each other on the day's findings. Normally scheduled at the end of the afternoon, the "five o'clock briefings" became an ideal way to swap information, review monitoring problems and double-check radiological findings. When a team reported an unusual set of numbers, readings that did not correlate with other results, briefing participants would go to large maps with sampling locations plotted on an overlay. Rechecking the information with other teams which had worked in the area always solved the anomaly, usually the result of an error in transposing the readings. This peer review quickly corrected any inaccurate or hasty calculations. At the end of the meeting, command post staff would collect, collate and distribute the environmental data to the Nuclear Regulatory Commission in Bethesda and to Pennsylvania officials at the Fulton Building in Harrisburg.[5]

In spite of the five o'clock briefings, the peer reviews and the dissemination of the data, one significant problem remained. However often the data were checked and rechecked for radiological accuracy, no one at the command post — neither the radiological assistance teams nor the federal agency representatives — was certain that the information was being used by either the Nuclear Regulatory Commission or Pennsylvania agencies in making decisions. No one was certain, for example, whether the command post environmental monitoring data had been considered in preparing the Governor's evacuation advisory. Dr. Walter H. Weyzen of the Department of Energy had spoken with state officials in Harrisburg and had complained Sunday that monitoring information was not being utilized. Scientists, too, became puzzled over the complaint. No radiation levels they had seen warranted evacuation, and some began to wonder if the Governor or the Nuclear Regulatory Commission knew something they did not know. Since communication between the command post and the Unit 2 control room was nearly non-existent, this was a possibility.[6]

Residents continued to fear the reactor. Some fled the area, while others took advantage of the situation. Upon certification by the Red Cross, individuals claiming to be displaced due to the "reactor situation" were given five hundred dollars from insurance companies to aid in relocation. According to the local director of the Red Cross, many were taking the money and then going over to the Red Cross shelters in Hershey.[7]

The probability of public alarm from the stricken plant was greatly reduced by April 4, one week after the accident had begun. Radiation levels during periods of planned venting were much lower than readings taken previously during similar operations. Dosimeters were measuring steadily decreasing dosage rates in the towns around Three Mile Island. Concentrations of radioactive gases in the containment building had also experienced a marked decrease. The Food and Drug Administration did report finding very low levels of radioiodine in some milk samples and in samples of wastewater dumped from the plant into the Susquehanna River. This caused some concern, not because of the levels, which were far below the Environmental Protection Agency's accepted concentration limits, but because officials did not know the source of the contamination. Even so, the Department of Energy sent copies of the booklet, "Emergency Handling of Radiation Accident Cases," to Pennsylvania officials in case the situation worsened.[8]

A preliminary assessment of the impact on the health of the population within fifty miles of the plant was completed on April 3. Drawn up by a team of specialists from the Department of Health, Education and Welfare, the Environmental Protection Agency and the Nuclear Regulatory Commission, the report estimated the number of fatal cancers which might occur due to the accident at Three Mile Island. The answer: .36. Cancer and genetic damage, the report continued, statistically would affect only one person. The Nuclear Regulatory Commission chose not to release the estimates to the press.[9]

By April 3, Department of Energy teams had analyzed 165 vegetation samples, 74 water samples, 134 soil samples and numerous air samples. Of these, three gave "slight positive identifications." All other samples indicated [normal levels of] background [radiation]." In short, the scientists had not yet found any significant public health threat.[10]

☆ ☆ ☆ ☆ ☆ ☆ ☆ ☆ ☆ ☆ ☆ ☆ ☆

Initially, the fact that no unusual levels of radiation had been detected by the Department of Energy's radiation assistance teams had little effect upon the Department's response. One week after the accident had begun, more than one hundred Department of Energy personnel at Three Mile Island were not as concerned by their own data as by the public's perceptions and fears of the twelve hundred millirem per hour radiation release, the apparently explosive hydrogen bubble, the Governor's evacuation advisory, and the general tension generated by the media. The Department thus brought additional resources to the scene, partially to assure state officials and the Nuclear Regulatory Commission that large numbers of experts were working on the problem or standing by should they be needed. But because of the Department's low profile, the extent of its response was not made public; consequently, people living near the plant remained largely unaware of the resources sent to protect them.

Yet even as more and more radiological assistance teams and equipment arrived, most of Friday's problems had subsided. No more major radiation releases were reported. The hydrogen bubble had mysteriously disappeared. There were more troops on hand than the battle warranted or, perhaps, the generals could efficiently direct.[11]

Given the low radiation levels, scientists on the radiological assistance teams became convinced that the emergency phase of the operation was over and the monitoring teams should go home. Bob Friess reported that his Brookhaven team members were finally getting enough rest, a sure sign that the crisis was passing. Although some iodine samples indicating radioactivity above the minimum detectable level were taken, the concentration was nowhere near the danger level, and scientists agreed that environmental monitoring could be concluded. Moreover, the response had become so large that regular work at the laboratories was being neglected.[12]

Nonetheless, the monitoring operation was assuming an inertia of its own. Pennsylvania officials, happy with the work the radiological assistance teams were doing, wanted to keep them around, Dave Schweller felt, even though radiation levels were low. In addition, the Department of Energy teams possessed the best equipment and the longest radiological experience. So while nearly every scientist agreed that the response to Three Mile Island could be terminated, Schweller theorized that the response had become political, rather than radiological, and therefore far more difficult to bring to an end.[13]

Schweller was partially right. Politics did not care about low radiation readings; the political situation demanded a response and a presence. No one in charge of the Department's efforts, either in Washington, Germantown, Capital City Airport or the national laboratories wished to appear unresponsive. Furthermore, there was an additional reason to have the environmental monitoring teams at the scene. While the Interagency Radiological Assistance Plan outlined how a response might or would occur, nothing in the plan indicated the size of the response, or when and how a response should end. In that respect the plan was a nightmare of uncertainty to bureaucrats who favor clear responsibility on paper. Consequently, the operation was phased down on an *ad hoc* basis and those at the site had different perceptions of how and when to reduce the force than did those at the Emergency Operations Center.

Scientists working out of the command post believed that once a cold shutdown of the reactor was achieved, everyone could leave. Until then, most thought that the best plan was to maintain a reduced force to babysit the reactor. Friess noted that the twenty-seven contractor employees and two Department of Energy officials doing monitoring for the state, plus the Environmental Protection Agency's twenty-five people doing similar surveys, were more than adequate. Friess proposed that the Department lower its commitment to twelve, enough, he felt, to carry the workload. Furthermore, the Environmental Protection Agency could assume some of the responsibilities and monitoring functions of the radiological assistance teams from the Bettis and Knolls Atomic Power laboratories.[14]

At the Emergency Operations Center, Jack McCool already considered the number of monitoring personnel excessive. But he was uncertain how the situation could be remedied without prior approval from Pennsylvania or the Nuclear Regulatory Commission. Until the word came from higher up, McCool continued to rotate replacements into Three Mile Island, keeping personnel levels constant. McCool was willing to ask those in the field for suggestions on planning reduced levels of support, but whatever the final decision on a reduction in manpower, he was adamant that plans would be made by the Emergency Operations Center in consultation with the Nuclear Regulatory Commission and not by the people in the field.[15]

The persistent question of who was running the Department of Energy response strained the relationship between those at the command post and the Emergency Operations Center. Uncertainty about the seriousness of the accident led to different perceptions. The command post viewed the situation with less concern than Germantown. Scientists monitoring the environment came to regard officials not directly working at the site with some contempt. How could people tucked away in the basement of the Department of Energy building really understand what was happening at Three Mile Island?

The basis for this confusion, frustration and annoyance was communications. While telephone contacts between the Emergency Operations Center and the command post were difficult and irregular during the first days of the crisis, the real difficulty stemmed from a mutual misunderstanding of the pressures on those in the field and those in Germantown. So after the first week the communications problem was not technical, but organizational. Without prior experience at working together on a crisis of these dimensions, neither group fully understood the other's mission.

The Emergency Operations Center, fielding requests for assistance from the Nuclear Regulatory Commission and Metropolitan Edison, keenly conscious of Congressional concern and briefing anyone who telephoned the control center or crowded into the message center for the latest information, began to view the Department's response in terms of numbers. The more Department of Energy resources at the site or on alert, the more assurance and certainty in a time of instability from the point of view of the Emergency Operations Center and its Washington constituency. Thus local pressures determined level of response and determination to control the operation from Germantown.

Scientists in the field tended to be insensitive to Washington's political pressures. Furthermore, none of the radiological assistance teams came prepared for a continuing response. The teams from Brookhaven and Bettis rotated monitoring duties according to their own plans and did not always clear these arrangements with the Emergency Operations Center. When Bob Friess pulled the Brookhaven team out on the thirtieth of March, McCool said he was "just delighted with the service they gave us but we just can't have a dozen people dipping in and out at random." He wanted to coordinate through one channel "and that channel is from here to the site." In short, all decisions on assistance would be made in Germantown.[16]

This view eroded the initial idea of coordination between Germantown and the command post. The Emergency Operations Center realized that an on-site representative was necessary to coordinate the Department of Energy response properly. Thus Ed Patterson and then Joe Deal were dispatched to provide this coordination. But at the same time, the Emergency Operations Center was reluctant to surrender its own coordinating role. Therefore two levels of coordination emerged, one at Capital City Airport and the other in Germantown, each with its own perception of what was happening and what was needed.

Deal and others at the command post chafed under McCool's directions. "We think we can manage what has happened here without any problems," Deal had told the Emergency Operations Center on March 31. The Nuclear Regulatory Commission was discussing the possibility of an evacuation, Deal had noted, and he wanted to be certain that any evacuation of Department of Energy personnel would be handled from the command post, not Germantown. And he did not want any more environmental monitoring teams, Deal emphasized, as there was already a problem keeping those at the site busy.[17]

Indeed, even then Deal was experiencing difficulty finding any significant work for the experienced radiological teams that continued to stream into the area. When a group from the Mound Laboratory arrived on Saturday, March 31, from Miamisburg, Ohio, Jim Sage of Bettis briefed them on the situation, but he had nothing for them to do. In a cold rain they drove down to the Fulton Building in Harrisburg to see if they might help the Food and Drug Administration crews. The Food and Drug Administration was in the process of placing dosimeters in the outlying areas. They welcomed Mound's offer. Ironically, neither the people from Mound nor the Food and Drug Administration knew the area well, which delayed placement of the dosimeters. Mound scientists were pleased to be helping, but they felt frustrated. Trained health physicists and nuclear technicians, they had believed they were driving all the way across Ohio and Pennsylvania to monitor radiation. Taping dosimeters to churches and fire stations was, they felt, a gross misuse of their expertise.[18]

Nonetheless, the number of Department of Energy people coming into Three Mile Island continued to expand. On April 5 there were 102 Department and contractor employees operating out of the command post. By then, most of the Department's national laboratories were involved with Three Mile Island, including the Bettis and Knolls Atomic Power laboratories, facilities which had had long experience with pressurized water reactors. Both laboratories had been established in the early stages of the naval program headed by

Admiral Hyman Rickover, who had pushed the development of the pressurized water reactor. Other scientists from Oak Ridge and Argonne and the Department's Environmental Monitoring Laboratory went "plume chasing" to determine if the invisible cloud contained dangerous radioactivity. After receiving plume predictions from the ARAC system, whose computer was in California, the scientists would pile into a "chase" car to test the region where the plume bumped the ground. There the teams would take soil, water and vegetation samples, analyze them in a mobile laboratory, usually a specially equipped van or recreational vehicle, and then exchange the information at the five o'clock briefings.[19]

The team from Oak Ridge, which arrived on March 30, worked for the Nuclear Regulatory Commission out of its command post at Trailer City. Monitoring shifts were arranged to provide twenty-four-hour coverage of the area on the Middletown side of the river. The Argonne radiological assistance team, also operating out of Trailer City, surveyed the west area of the river, around Goldsboro.[20]

The monitoring teams from each laboratory established sites at which they would install recording equipment or take samples on a scheduled basis. Most though not all of the sites were on lawns, in pastures and fields near roads. Some were placed on churches or fire houses. This made access to the instruments easier and it was thought that devices attached to buildings would not attract much attention. Moreover, they would be safer from vandalism.[21]

If the instruments were better guarded at a fire house, they were also closely scrutinized. Periodically, firefighters checked the monitors and compared readings with their counterparts in other towns. Firefighters in Newberry became agitated when they discovered that the readings on their fire house were much higher than those reported in Goldsboro. Only after a scientist explained that readings from two different kinds of instruments could not validly be compared did the anxiety lessen. Other firefighters were more confident of the situation from the beginning — or almost the beginning. The day after President Carter's visit, a Hershey firefighter was spotted wearing a T-shirt that proclaimed "I survived TMI."[22]

On Sunday, Idaho scientists Larry Ybarrondo and Nicholas Kaufman of EG&G flew to Three Mile Island at the invitation of Herman Dieckamp, the president of General Public Utilities, to take part in the think tank's deliberations. It was a rainy day, there was a good deal of confusion at the Air National Guard Armory in Middletown and they missed the first meeting of the Industry Advisory Group. The scientists reassembled on Monday and seven working groups were established to analyze the different parts of the reactor: the primary system, the secondary system, the residual heat system, degasification and cooling, core status, degradation scenarios (if "X" happens, what steps should be taken) and emergency power.[23]

For the first couple of days the teams met to discuss problems and advise the Nuclear Regulatory Commission. But little of their advice was having any effect, they learned. Kaufman was particularly frustrated, even angry, at the lack of any feedback. Others in the group announced they would quit and go home.[24]

The morale problem at the think tank as not solved until Wednesday, a week after the accident began, with the arrival of William Lee, the president of the Duke Power Company and a leading figure in the nuclear power industry. Lee reorganized the working groups, held daily status meetings and saw that the think tank's ideas and suggestions reached the Nuclear Regulatory Commission and Babcock and Wilcox. Nevertheless, the crisis was by then over and the industry's think tank had failed to make the impact it might otherwise have enjoyed due to the initial shoddy organization and poor communications.[25]

While the scientists worked on the semiscale simulation tests in Idaho, some non-scientists claimed to know why the accident had occurred. On April 2, security officers of the United States Labor Party told Federal Bureau of Investigation agents that the incident was caused by sabotage. The accident, the Party officials assured the agents, was the result of a conspiracy composed of the producers of the film *The China Syndrome*, Secretary of Energy James R. Schlesinger, National Security Advisor Zbigniew Brzezinski and former Secretary of State Henry Kissinger. The agent dryly noted in his report that the Party's security officers offered no proof of sabotage and that there was "no substance" to the allegation.[26]

☆ ☆ ☆ ☆ ☆ ☆ ☆ ☆ ☆ ☆ ☆ ☆ ☆ ☆ ☆ ☆

In the final analysis, it became clear that there was only so much that could be accomplished at Three Mile Island by the hundreds of experts responding to the accident. The small amount of radioactivity escaping from the plant could be adequately monitored by fewer people. On April 6 the planned reduction of the Department of Energy's activities and manpower at the site began. Ed Patterson headed the planning effort for the reduction and proposed to cut the number of people at the command post to forty-two by the end of the week. Should a problem arise, Patterson would keep an additional forty-eight people at the national laboratories on a six-hour alert. Patterson thought that this plan would provide adequate support until the reactor had reached a cold shutdown state, when the heat from the core had dropped to the point where the coolant would not boil.[27]

Even as Patterson was presenting his plan to reduce the level of manpower, radiological monitoring operations were already being reduced. The Nuclear Regulatory Commission had cancelled the midnight and three a.m. AMS/NEST flights, ARAC had decreased the number of its plume predictions and the ground monitoring teams were cutting back the frequency of their surveys. Neither Thomas Gerusky nor Bernie Weiss, who was directing the Nuclear Regulatory Commission's efforts from Bethesda, had any objection to the plan to reduce manpower. But all parties wanted to be certain that both Harold Denton and top officials of the Department of Energy okayed the plan. By the end of the week all the necessary approvals had been received and most of the monitoring teams had gone home.[28]

Even as the plan was first being discussed at the command post and the Emergency Operations Center, participants at a meeting in Washington were also suggesting a significant reduction in the Department of Energy effort. Several days after the beginning of the accident, senior officials at the Department of Health, Education and Welfare and the Environmental Protection Agency had realized that the data going to the Nuclear Regulatory Commission were being collected by the Department of Energy. Furthermore, they had discovered that the Commission and Metropolitan Edison had very little independent monitoring capability. "There was very much an unspoken feeling that . . . there ought to be independent radiation monitoring up there," recalled Steven Gage of the Environmental Protection Agency. Environmental data should not be "tied with what the . . . public might characterize as a pro-nuclear cabal," he told the Kemeny Commission. Consequently, both agencies sought to become involved in monitoring the accident. No one at the meeting, apparently, considered how the monitoring information was being used by the Commission or the utility; they were concerned primarily with the public's acceptance of the information, not its application at Three Mile Island.[29]

At the Environmental Protection Agency's offices at Waterside Mall in southwest Washington, officials catalogued their monitoring capabilities. The Agency's Office of Radiation Protection could commit eight to ten people and a mobile van with "a very limited amount of radiation monitoring equipment." The Agency could also send an airplane and thirty monitoring stations from Las Vegas, where the Agency had had experience surveying the weapons test site. A small crew from the Department of Health, Education and Welfare's Food and Drug Administration were already taking milk samples in the Harrisburg area.[30]

Officials from both agencies became alarmed when they discovered at the March 31 White House meeting that the Department of Energy was not only collecting environmental data at Three Mile Island but was coordinating all off-site monitoring activities. The officials assumed that Congress and the public would not believe that "the data was [sic] being collected carefully and objectively," although there was no evidence to support this assumption, as Richard Cotton, an assistant to Secretary Joseph Califano of the Department of Health, Education and Welfare admitted. Cotton believed that the Environmental Protection Agency should be the dominant coordinating agency. But his experts at the site told him that the Department of Energy's monitoring capabilities were crucial. As a result, when Cotton and Gage sat down on April 10 to draft a memo for White House assistant Jack Watson's signature designating the Environmental Protection Agency as the main federal agency to monitor radioactivity at Three Mile Island, they specifically assigned certain monitoring roles to the Department of Energy.[31]

The Department of Energy's participation was probably assured after the visit of Cotton's "expert," John Villforth, to the command post on April 6. Villforth arrived to tour the facility and immediately announced that someone else would be taking over for the Department of Energy. Joe Deal, the senior officer in charge, was at once surprised and amused by Villforth's pronouncement. Facetiously, Deal said the Department could move its resources out by nightfall and Villforth could move his in. Realizing the paucity of Health, Education and Welfare's monitoring assets, Villforth reportedly backed down, but Deal thought it would be best to get the working arrangements down on paper.[32]

Deal needed an official paper, so Thomas Gerusky obliged. As director of the Pennsylvania Bureau of Radiation Protection he wrote Deal a letter requesting that the Department of Energy "collect and collate all environmental sampling data from all Federal agencies and the State of Pennsylvania in the Three Mile Island area." The Department's efforts, he wrote, "have been exceptional."[33]

Gerusky's memo had no effect in Washington, where the Gage-Cotton memo received Jack Watson's blessing. A Watson aide phoned Secretary Schlesinger to see if he had any "difficulty" with its contents. Schlesinger felt that the Department of Energy could perform the primary agency role, but he had no objections. So on April 13, a week after the Department had already begun to cut back its resources, the Environmental Protection Agency took over the long-term monitoring at Three Mile Island.[34]

By Wednesday, April 11, two weeks after Unit 2 had scrammed, the environmental monitoring forces had dropped to fifty-eight people. Easter and income tax demanded more attention from many of the scientists, who wished to spend the holiday with their families or to make last-minute scrambles to give Uncle Sam his due. As Easter approached, more and more people went home. Denton insisted that monitoring continue through Monday the 16th, the day after Easter and the last day for mailing tax returns. He wanted to check on any problems that might arise until a cold shutdown could be achieved.[35]

On April 16, Joe Deal met with Denton and other Nuclear Regulatory Commission officials to reach a decision on the continuing levels of environmental support. From that meeting and a later conference with Colonel Oran Henderson of the Pennsylvania Emergency Management Agency and Gerusky, Deal and the others agreed that the AMS/NEST helicopter would abandon its scheduled flights and fly only upon request. ARAC would continue to furnish appropriate meteorological data. All ground monitoring teams from the national laboratories would be released to go home, and any additional sampling would be done by the Department of Energy's Environmental Monitoring Laboratory. By the beginning of May, the Environmental Monitoring Laboratory had arranged to continue long-term soil and vegetation measurements at six to eight locations, reporting their findings to the Environmental Protection Agency.[36]

Even before most monitoring teams had left the Three Mile Island region, they had taken and analyzed over nine hundred examples of vegetation, soil, water and air. In addition, nearly four thousand instrument readings had been made as the crews drove from one location to another. AMS/NEST had flown over sixty aerial monitoring missions, and the Nuclear Regulatory Commission requested that the helicopter and crew remain at the Capital City Airport until June. The Commission also asked the Department of Energy to refly the 1976 background survey of the area to determine if any detectable changes in radioactivity had occurred as a result of the accident.[37]

At an April 24 meeting of federal participants in post-accident monitoring, the Nuclear Regulatory Commission, the Department of Health, Education and Welfare, and the Department of Energy agreed to exchange pertinent data regularly and continue to release this information through the Nuclear Regulatory Commission. The Environmental Protection Agency refused to sign the agreement, which might have threatened its position as the dominant agency, a position awarded by the White House two weeks before. Interagency rivalry was still a part of the political response.[38]

For the duration of the Department of Energy's environmental monitoring presence at Three Mile Island the samples analyzed by the scientists were, by and large, negative. No radioactivity threatened public health and safety. Aerial measurements did reveal the presence of xenon from the reactor and the ground teams did find traces of radioactive iodine. In addition, the sensitive instruments also discovered the presence of cesium in a few soil samples. However, it had not come from Three Mile Island but from the Chinese nuclear test fallout in the autumn of 1976. Indeed, the extensive data base generated by the monitoring surveys during the twenty days following the accident documented clearly the absence of any radiological hazard outside the plant to the public.[39]

☆ ☆ ☆ ☆ ☆ ☆ ☆ ☆ ☆ ☆ ☆ ☆ ☆ ☆ ☆ ☆

While the radiological assistance teams and the AMS/NEST helicopter continued to check the environment for hazardous radioactivity outside the power plant, others began to focus their attention on the fission products trapped on the island. Communications between the utility's personnel in the control room and the command post had never been good. The fact that an overzealous plant security guard had threatened to shoot down the Depart-

NUMBER OF PLUME FLIGHTS

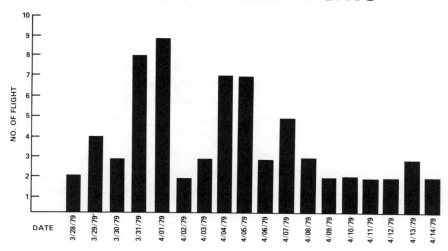

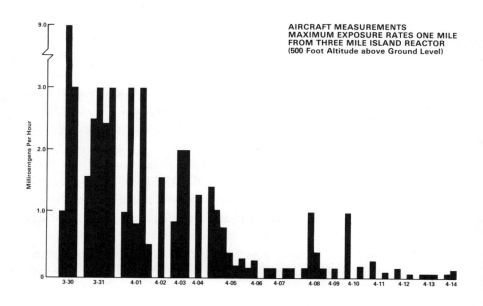

AIRCRAFT MEASUREMENTS
MAXIMUM EXPOSURE RATES ONE MILE
FROM THREE MILE ISLAND REACTOR
(500 Foot Altitude above Ground Level)

DOE — CONTRACTOR PERSONNEL

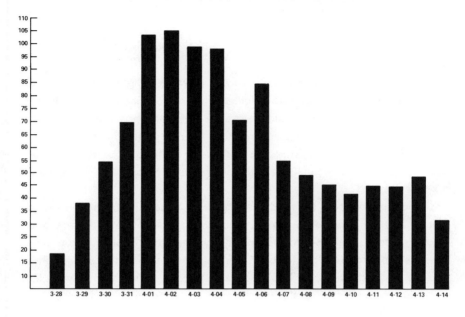

ment of Energy's helicopter if it flew too close to the plant had not improved the situation. Yet it was crucial for the command post to get some kind of information about the situation at the plant, especially when the Nuclear Regulatory Commission requested a special helicopter flight or transportation for a core coolant sample.[40]

Usually even the basic information for coordinating flights with activity on the island was lacking. Often the Nuclear Regulatory Commission representative in the control room would phone the command post and ask that the AMS/NEST helicopter go up and monitor the radiation from a venting that would shortly take place. More than a few times the pilot and crew would stand by for hours or actually be in the air only to learn that the release would not take place.[41]

The command post believed that the only accurate information on the reactor and the plant came from Harold A. (Hap) Lamonds, who managed EG&G's program in Nevada. Lamonds had taught Harold Denton some years before when Denton had been a young student at North Carolina State, and he also administered licensing examinations to nuclear power plant operators on the West Coast. Possessing many contacts with the Nuclear Regulatory Commisson and utility company officials, Lamonds moved easily from the control room of Unit 2 to the think tank to the command post, where he would show up for the five o'clock briefing. His information did help the AMS/NEST crews and the ground teams get a handle on what they might expect to "see" with their instruments.[42]

No one, however, knew yet exactly what had happened inside the reactor. One way of getting some idea was to analyze the contents of a sample of coolant — demineralized water which circulated around the fuel rods and cooled them — from the reactor core. Under pressure and super-heated, this water would travel through a series of pipes into a heat exchanger, where it would heat more water to drive the steam generators. The coolant swirling around the uranium fuel rods was radioactive. If there had been any damage to the fuel rods when the core was uncovered, the detritus from the damage would be carried in this primary coolant. Thus, coolant analysis would indicate the extent of damage to the fuel rods and help identify the fission products present.

Because of the radioactivity, the primary coolant samples had to be handled with great care. Although Herman the Robot was not used, a number of samples were taken. Drawn off into a metal container resembling a thermos bottle, the samples were then placed inside a specially designed fifty-five gallon drum and driven to Capital City Airport. There the drums were loaded on a giant C-141 military transport and flown to the Bettis Laboratory in Pittsburgh for analysis. Later samples showed that while the reactor core was indeed damaged, very little fuel melting had occurred.[43]

The analyses of the primary core coolant samples indicated a change in concern from immediate problems, such as a hydrogen bubble and radiation leaks, to investigating the causes of the accident and the long-range planning necessary to recover from it. Seven days after the accident had begun officials believed that the reactor had stabilized and that the threat of any immediate danger off site had disappeared. At that point the Nuclear Regulatory Commission and the Metropolitan Edison Company turned again to the Department of Energy for planning and advice on the decontamination procedure.

For this second phase in the effort to conquer the lingering problems caused by Three Mile Island, different troops were needed. The shock troops of the radiological assistance teams, those prepared to respond to emergencies and environmental threats, gave way to crews experienced in nuclear decontamination who would literally mop up the radiation on the site, wearing special uniforms and masks and swabbing the floors with long-handled mops and buckets of soapy water.

On the evening of April 1, hours after President Carter had left Three Mile Island, Lee Gossick, the Nuclear Regulatory Commission's Executive Director of Operations, called the Emergency Operations Center. It was urgent that the auxiliary building be decontaminated if the reactor were to reach cold shutdown, he said. Moreover, the Commission and the utility wanted to start the process the next morning. About three hundred people would be needed for the operation and Gossick wanted the Department of Energy to handle the logistics. At about the same time, Walter Creitz, the president of Metropolitan Edison, phoned Joe Deal at the command post with a similar request.[44]

Speaking with Germantown about the request, Deal promised to meet with Gossick at noon the next day to review the situation. But any decision on decontamination, Deal pointed out, would have to come from other Department officials. In the meantime, Deal would see what was needed. He asked that the Emergency Operations Center send Ed Vallario to Harrisburg for the noon meeting, as he wanted him to draw up a plan for re-entering the building. Vallario was no stranger to damaged reactor areas. In January 1961, he and another man had gone into the highly contaminated reactor building at the SL-1 facility in Idaho in a vain attempt to rescue three men from the reactor core area after radiation had escaped from a control-rod aperture. Deal wanted Vallario to write up a report on decontamination procedures.[45]

Vallario, Deal and other Department of Energy representatives met with Herman Dieckamp of General Public Utilities on April 5. The auxiliary building was the utility's top priority, Dieckamp told them. Decontamination of the containment building was a long-range problem. In both cases, General Public Utilities would need the Department of Energy's help because the Department had considerable decontamination experience. But Dieckamp could not be any more specific. Deal suggested that as recovery operations progressed, the utility might pinpoint specific areas where the Department could be of help. Then, Deal said, the matter might be discussed more formally.[46]

One reason for Deal's reluctance to become involved in the decontamination process was that the responsibility and expertise within the Department of Energy resided with Energy Technology, not Deal's environmental branch. Within Energy Technology was the Office of Nuclear Energy Programs and that office had from the beginning viewed the Three Mile Island accident from the point of view of decontamination. At the March 31 White House meeting, Energy Technology representatives Jack Crawford and Herb Feinroth had remained silent, unaware of the Department effort at Three Mile Island. When the meeting had ended, Crawford had gone to Joseph Hendrie and assured him that the Department of Energy was ready to assist at Three Mile Island, especially in handling large volumes of radioactive materials and in cleaning up contaminated areas.[47] A week after the meeting it was clear that such assistance was essential.

Nonetheless, if there was confusion within the Department about the nature of the federal response, the Office of Energy Technology was vitally interested in the short and long term implications of the accident. Its scientists knew, for example, that at least since November 1977, Babcock & Wilcox plants had had problems maintaining pressurizer water levels during reactor trips, especially during a loss of feedwater into the primary coolant system. For the first several days after Feinroth's aborted trip to act as the office's site representative, the Nuclear Programs officials watched the events unfold from Germantown. In the short term they were concerned with the damage to the core and finding what had caused the accident. From a longer perspective, they studied the accident in terms of decontamination and the financial problems that would arise.[48]

Still, from the first, decontamination occupied a prominent place. On April 2, Feinroth suggested that the DuPont Corporation, which ran the Department of Energy's facility at Savannah River, Georgia, could supply appropriate personnel to clean up the plant. DuPont was experienced in planning, managing and carrying out major decontamination projects, Feinroth pointed out, recalling that DuPont people had cleaned up after the weapons accidents at Palomares in Spain and Thule, Greenland.[49] The Nuclear Regulatory Commission was also looking ahead, wondering what could be done with the damaged reactor core once the plant had been decontaminated. Initially, they hoped to send it to the Nevada Test Site, but other options, such as Idaho, were also open.[50]

Concern about disposition of the reactor core was premature. First the site had to be decontaminated so people could work inside the buildings. As in the initial response to the accident, dozens of experts were brought in from across the nation. Organization of the post-accident response was carried out on an *ad hoc* basis. No one appeared to know who would direct the clean-up operation. Part of the confusion arose because of the number of different organizations requesting the assistance. The Metropolitan Edison Company, the Nuclear Regulatory Commission and the Electric Power Research Institute all invited specialists to work on the broad variety of problems the accident presented. Few of the requests, even for Department employees, came through the Department of Energy's Emergency Operations Center. In fact, the only way the Department learned of the arrival of other scientists' to work at Three Mile Island came if they happened to stop by the command post.

The Emergency Operations Center was aware of this changing of the guard and helpless to direct it. The separate mission to assist in the cold shutdown and decontamination process was moving beyond the planning stage and neither the Office of Environmental Compliance nor the Emergency Operations Center would play much of a role. When Willis Bixby, a Department of Energy scientist from the Idaho National Engineering Laboratory, arrived in Middletown on April 8, the command post and the Emergency Operations Center were a bit surprised. The Nuclear Regulatory Commission had asked Bixby to coordinate the work of the various laboratories analyzing the reactor's problems. Bixby checked into the command post to attend one of the five o'clock briefings in the hangar. Involved almost exclusively with the monitoring effort, the command post suggested that Bixby talk with Eric Beckjord in the Office of Energy Technology, who was running the long-term recovery phase. "It looks more and more like there's need for a new organization up here," Ed Patterson told Germantown, "and Bixby may be the forerunner of it."[51]

As it turned out, Bixby was in fact the forerunner of a new organization at Three Mile Island, one which would be directed through the Office of Energy Technology. A week after the accident, on April 5, when it was clear to most people at the Department of Energy that the emergency phase was drawing to a close, Assistant Secretary for Energy Technology, Robert L. Ferguson, and Andrew J. Pressesky of Ferguson's staff drove with Beckjord and Feinroth to Three Mile Island for a first-hand report. They toured the command post, where Joe Deal briefed them on the status of the environmental monitoring. They also met with Harold Denton to learn the status of the plant, with William Lee of the Duke Power Company and with Herman Dieckamp of General Public Utilities. From Lee and Dieckamp they heard that the power companies had called in some of the most experienced nuclear reactor scientists in the country to study the reactor and decide how the various problems should be handled. Finally they spoke with Milt Levinson, who was directing the Industry Advisory Group's think tank. Levinson believed that the reactor core had been uncovered three separate times on the twenty-eighth and that significant damage had occurred in the core, an assessment that varied from the inital reports on core damage from the Bettis Laboratory.[52]

Touring the area with Ferguson, Pressesky and Beckjord were two others who would become involved with the Department's post-accident response. Richard G. Hewlett and Jack M. Holl of the Historian's Office had made the trip at the request of Joe Deal and Herb Hahn. During the first days of the accident, Under Secretary Dale Myers had mentioned the importance of capturing the historical record of the accident. But the historians, more comfortable in the past than in the present, had taken nearly a week to get to Harrisburg. Once there, however, they spoke with Deal, Hahn and others at the command post, urging them to collect and save the materials which documented their monitoring activities. Not only would this material prove important for historians but it would be invaluable to the participants in analyzing the Department's response. [53]

The visitors were impressed by what they heard and saw. But it was clear that a successful recovery operation was crucial to the future of the commercial nuclear power industry. Whatever lessons could be learned from Three Mile Island and applied to other nuclear programs within the Department of Energy, they believed, would be highly beneficial both to the federal government and the energy future of the nation. The visitors also realized that the Energy Technology presence would be a long one, lasting while the recovery phase continued. Neither the Nuclear Regulatory Commission nor Metropolitan Edison had any specific plans for Department of Energy involvement, so Ferguson decided to establish a liaison between the Commission and his office. Beckjord and Pressesky initiated meetings with the Nuclear Regulatory Commission and the utility on April 11 to discuss future assistance. But rather than coordinate their activities through the airport command post, which was eight miles from the reactor, they decided to locate the Energy Technology liaison representative in one of the Nuclear Regulatory Commission's trailers in Trailer City. [54]

By May the trips to Trailer City were being made on a regular basis. By talking to the people working at the site on the recovery of the plant, the Office of Energy Technology believed it could better decide what kind of technical or financial assistance could be suggested or given. The assumption was that the size and seriousness of the Three Mile Island accident would test recovery procedures which could be applied to other Department of Energy nuclear programs. [55]

In late May, Feinroth's office was providing assistance to General Public Utilities' efforts to clean up the contaminated water at the site. The office also expressed an interest in evaluating the damaged reactor core, although any analysis could be, according to many estimates, five years away. [56]

With the focus on such long-range planning and the fact that little was being done which actively involved the Department of Energy, the link with the Nuclear Regulatory Commission at Trailer City soon reached "the point of diminishing returns," according to one of the representatives. All that was really necessary was that the Commission's officials at Three Mile Island send a daily report to the Office of Energy Technology and that someone call Robert

Arnold, General Public Utilities' vice-president for operations, several times a week to get the latest information on company plans and activities. But in any case, the duties of the liaison representative were not clearly defined. The conditions at Three Mile Island, it was noted, were unlikely to require Department of Energy assistance as long as the approval to move contaminated water was delayed, and in the meantime the assignments to Three Mile Island would be "lengthy and unrewarding." Only when there was significant activity should someone be on the site.[57]

Yet Department scientists trekking to Three Mile Island did not make policy and the liaison program continued through the summer. One nuclear expert who made the trip defined the Three Mile Island accident as a "relatively minor incident." But it was evident that the accident would have a not-so-minor impact on the commercial nuclear power industry. The nuclear branch of the Office of Energy Technology strongly believed that American reliance on foreign oil and the resultant shortages then occurring made it imperative that the office support any industry efforts to improve the operation and safety of nuclear power. Accordingly, decontamination planning among government and utility officials continued throughout the summer and early fall.[58]

In August, officials from the Office of Nuclear Technology met with their General Public Utilities counterparts to discuss problems that might be solved to the benefit of the Department, the utility and the nuclear power industry as a whole. In short, it was thought that information about the accident and recovery operations could be gathered, disseminated and applied throughout the country. For example, the Department of Energy wanted to know how the instruments and electrical components performed during and after the accident, how fission products were dispersed within and outside of containment in order to improve venting designs, and how to find innovative ways to decontaminate the Unit 2 containment building. The meeting also discussed the less technical — but far more important political — issue of removing, packaging, shipping and disposing of the damaged fuel core, as well as reducing the volume of radioactive wastes on the island.[59]

Personnel changes at the Nuclear Regulatory Commission and sweeping reforms in the Nuclear Regulatory Commission's organizational structure recommended by the Kemeny Commission, as well as opposition to nuclear power caused by the Three Mile Island accident, delayed any substantive action on decontamination. In mid-November 1979, General Public Utilities asked the Nuclear Regulatory Commission for permission to conduct a controlled purging of radioactive krypton gas from the containment building, a necessary step in recovering the reactor system and core. In February 1980, the Commission still had the request under "active consideration."[60]

The Department of Energy felt time was running out. The containment building had been isolated for over ten months, G. W. Cunningham, the new Assistant Secretary for Nuclear Energy, wrote to the Nuclear Regulatory Commission in February that further delay might be hazardous. The instruments in the containment building had not been designed to withstand the extreme conditions of 100% humidity which existed there. Many instruments needed to be replaced, one safety study urged, otherwise the risk to the safety of workers and the public would be increased. The cooling system, the study also pointed out, had been operating unattended for this entire period and if it failed, an uncontrolled release of krypton would occur. Yet as of May 1980 technicians still were unable to enter the containment building, because of a rusted door lock.[61]

Based on this study, Cunningham proposed that the best solution was to act promptly and conduct a controlled purging of the gas from containment when the weather conditions were most favorable. The Department of Energy, he assured the Nuclear Regulatory Commission, would monitor the environmental conditions off-site to insure that the releases were within acceptable limits.[62]

The guard at Three Mile Island had indeed changed. The emergency shock troops had given way to a much smaller army of occupation. The enemy had been contained, if not conquered. Throughout 1980 the army of occupation remained and would continue to do so until both the technological and political battles were over.

Chapter VII
Armistice

I don't see how you can sit there in Washington and imagine a situation like that if you had all the information in the world, because there are too many local things that affect what's going on.

L. Joe Deal

O ver a year later, Three Mile Island remains a combination battlefield and bivouac area. Workmen are repairing the wooden bridge leading to the southern end of the island. Two guarded checkpoints limit access, first to the island, then to the immediate area of the plant. Driving past the first guardhouse one is struck by the juxtaposition of man and nature. Piles of scrap metal lie rusting along the road. Nearby in a pond several mallard ducks and a solitary white American egret search for food. Suddenly a huge parking lot appears, as jammed with cars and pickup trucks as a busy shopping mall on a Saturday afternoon. The crisis of Three Mile Island is not over. Hundreds of men and women still occupy the island, working to contain the radioactivity inside the plant. An armistice is in effect on Three Mile Island. The initial battleground is relatively quiet. The principal battle has moved elsewhere.

The Three Mile Island accident has raised many issues concerning the past, present and future of nuclear power. What improvements should be made in reactor safety and design? In operator training? How should utilities insure themselves against major economic loss in the event of an accident? What measures should be taken to protect the public health and safety? To provide for rapid and safe evacuation from a hazardous area? How far away from densely populated areas should reactors be located?

For the Department of Energy there was an even more germane question: what lessons could be learned from Three Mile Island about the role of the Department in responding to civilian radiological emergencies?

The Department began its analysis immediately after the accident. On April 14 it issued a report on its radiological response to the accident which detailed the actions taken, equipment used and radiation monitoring results. On May 10 Jim Sage of the Bettis Laboratory concluded that the Interagency Radiological Assistance Plan was "very responsive and highly effective," and the command post team "functioned extremely well," but "had the need for a full-scale evacuation, major decontamination, or major construction been required, that *ad hoc* management group would have been severely taxed." On the matter of public information, Sage concluded that "the public health and safety profile was much too low. The lack of an adequate on-scene command structure by NRC blocked the flow of health and safety information from DOE to the NRC PIO [Public Information Office] staff."[1]

On May 10, 1979, an *ad hoc* Population Dose Assessment Group consisting of representatives of the Nuclear Regulatory Commission, the Department of Health, Education and Welfare, and the Environmental Protection Agency issued a preliminary report on the impact of the Three Mile Island accident on the population's health. Although the Department of Energy was not represented in this group, Department aerial and terrestrial monitoring data were used. The report concluded that the average dose to a person within a fifty-mile radius of the plant from March 31 to April 7, 1979, was 1.5 millirem. Appendices described the Department of Energy radiological measurements, their results and their analysis by Department laboratories.[2]

On May 15 and 16, 1979, a meeting was held at Department headquarters in Germantown to evaluate the response to the Three Mile Island accident. In general, those present felt that the response had been excellent in terms of technical data coordination and provision of communication facilities for state and other federal agencies. This was the first time a Department of Energy response had involved both field and headquarters staff, the Aerial Measuring Systems, the Atmospheric Release Advisory Capability, Radiological Assistance Teams from several regions, state authorities and federal agency members of the Interagency Radiological Assistance Plan.

Admittedly, there had been problems. At first it had not been clear who was in charge of the response, nor was it clear whether the Interagency Radiological Assistance Plan had ever been officially used. AMS/NEST communication equipment had been insufficient initially because the local telephone system was overloaded, and it was agreed that in the future communications systems should be dispatched to the site immediately. "The problems associated with communications," it was observed, "arose primarily from the hesitancy of the utility, the state, or the Nuclear Regulatory Commission to call for assistance."[3]

Other problems had cropped up as well. Groups and individuals stayed longer than anticipated, leading to fatigue, a lack of money, personal inconvenience and problems with motels and transportation. Radiological assistance teams needed better training on communication systems and environmental sampling. The AMS/NEST helicopter instruments were so sensitive that they were saturated by the plume; a broader range of detector sensitivities would have been useful. There was no plan to deal with public relations; silence and isolation made scientific operations more efficient, but releasing more information might have helped alleviate public concern.

Oak Ridge provided its own evaluation. These scientists felt that the Interagency Radiological Assistance Plan had never been activated and that the Department of Energy response was "unilateral." The Department mission was never clearly identified, nor was the chain of command from headquarters to the scene of the accident defined. Technical on-site support to the Nuclear Regulatory Commission and Metropolitan Edison was not coordinated, despite "major safety implications." Oak Ridge concluded that "there should be established an accident response management group at the HQ level consisting of representatives of all the involved federal agencies (with as IRAP is written, the coordinator being the Department of Energy)."[4]

In August 1979 the Nuclear Regulatory Commission's Office of Inspection and Enforcement published the results of its own investigation into the accident. A highly technical study, based on interviews with plant operators, the report concluded that despite some inadequacies in equipment, operator training and design, the accident "could have been prevented."[5] The report focused on the operations of Metropolitan Edison and had little to say about Department of Energy operations, since its account did not proceed in any detail beyond Thursday, March 29. In this version, the accident was a highly technical affair confined to the plant itself during the first two days, and the report attempted no interpretation of the accident in the context of any public, political or emergency response.

The President's Commission on the accident at Three Mile Island, chaired by John G. Kemeny, President of Dartmouth College, took testimony from two Department of Energy officials. On April 27 John Deutch, Assistant Secretary for Energy Technology, testified that the accident had been a "regulatory failure" to which the Department had responded with its radiation monitoring and laboratory analysis.[6] He could provide few details, however, since the main Department effort at Three Mile Island was not under his jurisdiction, but under the Assistant Secretary for Environment.

Perhaps with this in mind, the President's Commission heard comments on August 20, 1979, from L. Joe Deal, the senior Department official present at Three Mile Island. Deal reviewed various aspects of the Department's operations at the command post: its afternoon briefings for federal and state scientists, monitoring and data analysis, transportation of lead bricks for the recombiner, evacuation planning. Deal explained that, from a radiological standpoint, the major problem at Three Mile Island was the possible release of radioiodine, rather than the meltdown anticipated as a danger by some Nuclear Regulatory Commission spokesmen. Deal had no idea why the Department of Energy had not been invited to the White House meeting of agency representatives on Saturday, March 31. His main concern had been that radiological data gathered by the Department and given to the Nuclear Regulatory Commission did not appear to be getting back to Bethesda or Washington and having any effect on decisions there.[7]

In October 1979 the President's Commission submitted its final report to President Carter. The report found that at Three Mile Island the fundamental problems were "people-related problems and not equipment problems" and that "except for human failures, the major accident at Three Mile Island would have been a minor incident." It also found that the response to the emergency was "dominated by an atmosphere of almost total confusion" characterized by poor communications and lack of emergency preparedness.[8]

The President's Commission focused primarily on the activities of the Nuclear Regulatory Commission and the Department of Health, Education and Welfare, and had relatively little to say about Department of Energy operations at Three Mile Island. The Commission noted that Metropolitan Edison alerted the Department's Brookhaven office early on March 28, [incorrectly] that the Brookhaven team arrived at Three Mile Island at the request of the Nuclear Regulatory Commission and that Department helicopters were monitoring radiation by mid-afternoon on the first day of the accident. Further Department activity went unmentioned, except for an observation that in dealing with the hydrogen bubble the Nuclear Regulatory Commission "sought help from laboratories and scientists outside the Nuclear Regulatory Commission."[9]

The President's Commission made no mention of the Interagency Radiological Assistance Plan in connection with the emergency response. It recommended that responsibility for future emergency planning should rest with the Federal Emergency Management Agency, and that the Plan should be "reexamined and revised by the appropriate federal authorities in the light of the experience of the Three Mile Island accident, to provide for better coordination and more efficient federal support capability."[10]

An unpublished subcommittee report of the President's Commission did, however, deal with the question of the Plan and its effectiveness. The report found that Three Mile Island was, in effect, the "first test of IRAP in a response to a significant incident at an NRC-licensed nuclear power plant." Environmental Protection Agency and Health, Education and Welfare officials, however, "generally were unaware of IRAP's existence." The subcommittee made a distinction between the Department of Energy Radiological Assistance Plan teams at the national laboratories and the Plan. "Although IRAP signatory agencies provided indispensable technical monitoring assistance," the report noted, "much of this support was rendered on an *ad hoc* basis outside the formal structure of the plan." The Department of Energy was not mentioned.[11]

By the winter of 1979-1980 there was greater awareness that the Department of Energy had played a significant, if unrecognized, role at Three Mile Island. This was evident in the January 1980 report to the Nuclear Regulatory Commissioners by an independent investigating team headed by the Washington, D.C., law firm of Mitchell Rogovin. Despite the lack of any "designated lead responsibility for radiological monitoring and assessment," the Department of Energy effort had worked well, according to this report; the Department of Energy command post was efficient, and the Emergency Operations Center in Germantown, "intended primarily for weapons tests and accidents," had been useful because it could mobilize "more emergency monitoring equipment and other resources than any other agency."[12]

Rogovin investigators agreed with the President's Commission that federal emergency planning should be handled in the future by the Federal Emergency Management Agency. But it also recommended that the Department of Energy "be formally designated by executive order as the lead and coordinating federal agency to call upon and organize the emergency resources of all other federal agencies in case of an accident at a commercial nuclear plant requiring monitoring."[13]

☆ ☆ ☆ ☆ ☆ ☆ ☆ ☆ ☆ ☆ ☆ ☆ ☆ ☆ ☆

What is the larger significance of the Department of Energy's response to the Three Mile Island accident of 1979?

The accident now appears to have been more human drama than technological failure. Its principal actors were not scientists but plant operators, politicians and public officials. It was not so much a mechanical breakdown as a series of human choices that crippled a nuclear reactor and threatened public health and safety. The principal cause of the accident was human, not technical, error. From the outset plant operators chose to take actions that helped complicate a loss-of-coolant accident and release radiation from the plant. Federal and state officials then made numerous statements and recommendations, often on the basis of incomplete or inaccurate data, that were magnified by the media and transmitted to the world as the "Harrisburg Syndrome." In sum, Three Mile Island rapidly became a battleground because of what people feared might happen, rather than because of an accurate knowledge of what was actually happening. The major feature of Three Mile Island would appear to be a breakdown in human communications, not a breakdown of a nuclear reactor.

In the public mind Three Mile Island has nonetheless been widely perceived as a failure of nuclear technology. The accident is cited as an example of technocracy overriding democracy, of the failure of scientists to communicate risk to the public, of the end of a nuclear utopia of risk-free energy. In fact, nuclear power developed as a byproduct of the military weapons program, notably nuclear-powered submarines, and had been frequently oversold by its proponents in private industry as a safe and efficient energy source. From about 1966 to 1976, a decade of rapid growth in nuclear power, its proponents gradually shifted from trying to persuade industry that nuclear power was economical to trying to persuade the public that nuclear power was safe. By 1979, despite the rising costs of imported oil, neither industry nor the public was completely persuaded, and many orders for new plant construction were either cancelled or not placed. Even without Three Mile Island, the future of nuclear power was increasingly questionable.

In developing nuclear power, government and industry were by no means always in agreement. The utilities wished to develop a profitable nuclear alternative to fossil-fuel generating plants; government officials wished to protect the public from unnecessary risk by licensing and inspecting plants and enforcing their safe operation, and thereby resulting in a tug-of-war over nuclear power. By 1979 the tug-of-war involved four distinct elements: businessmen and industrialists associated with the power companies; local, state and federal officials concerned with the development of nuclear power plants; the antinuclear public-interest lobby or movement; scientists and engineers involved in the design of nuclear power plants and in protecting the public from radiation hazards. The Department of Energy effort at Three Mile Island drew primarily upon the fourth element, scientists from the national laboratories around the country. The more general response to the accident, of course, involved all four groups.

The division of authority in nuclear matters between the Nuclear Regulatory Commission and the Department of Energy in 1977 separated the regulatory and safety functions of the old Atomic Energy Commission. The health physicists and radiation experts available in a nuclear emergency thus operated under the environmental branch of the Department of Energy, or in the various national laboratories. These scientists had traditionally been concerned with the health and safety aspects of weapons testing. For them there was no particular commitment to nuclear power, which many viewed as an accidental byproduct of submarine propulsion, but there was a commitment to the radiological consequences of any potential threat to public threat and safety.[14]

This background is important in understanding that the Department of Energy's experience at Three Mile Island only partly reflected the larger pattern of events. Indeed, precisely the atypical and unknown aspects of the Department's response constitute its historical significance: the successful application of a technology of radiological monitoring, the flexible application of human decisions within a context of prior radiological experience and planning, and the great advantage of a quasi-military response to a civilian nuclear accident.

At Three Mile Island the new technology of radiological monitoring worked and worked well. The Brookhaven iodine filter enabled scientists to ascertain quickly that iodine releases, as distinct from xenon and krypton, were minimal. The ARAC computer system operated effectively in the first real test of its ability to forecast meteorological conditions in an area thousands of miles away from the computer itself. Thermal aerial photography proved able to identify reactor leaks. On-scene communications were greatly facilitated by the communications pods flown in from Las Vegas with packaged telecopiers, transmitters, and handi-talkies. The dedicated telephone lines of the command post were crucial in providing communication when local telephone systems were saturated. The Oak Ridge mobile manipulator, Herman, was available if needed. So were numerous devices crucial to logistics, supply and monitoring — helicopters, aircraft, radiation monitoring devices, maps and photographic equipment.

Yet what made the Department of Energy response most effective was the human component. The very same "human factor" that helped cause the accident became crucial in efforts to contain the crisis of Three Mile Island. From this perspective, the Interagency Radiological Assistance Program was less important than the longstanding practical experience of the radiological assistance teams that had responded over the years to other emergencies. Few scientists were actually aware of any plan, even though they were called to Three Mile Island by the Nuclear Regulatory Commission and Pennsylvania state health officials in accordance with the Plan's requirement; nor was the Plan ever activated by any single directive. That did not matter. What mattered was the prompt response of an army of professional scientists from the national laboratories who had had experience in previous nuclear tests and accidents. Many of these "old faces" knew each other well from laboratory training, weapons tests, study of fallout from the Marshall Islands, earlier emergencies or work for the Atomic Energy Commission.[15] The experience and know-how of these veterans often made communications easier, trust greater, response quicker and contingency planning better. What made the response effective was an existing cadre of professional radiation scientists who could be called upon in an emergency.

The human factor was therefore as important in helping contain the accident as in making it happen in the first place. Few Department of Energy scientists thought that in responding to the accident they were playing by the book; indeed, the experience of a nuclear power plant accident was for most of them unprecedented. Once called to the scene of the accident, many were forced to make choices that were *ad hoc* and unplanned, but important. When Bob Shipman did not know whether "Nine Mile Island" referred to

Three Mile Island or Nine Mile Point Station he decided to bring one hundred maps of two different reactor sites, rather than fifty maps of one. Herb Hahn chose to bring U.S. Air Force helicopters from Andrews Air Force Base to provide support, to set up command post at Capital City Airport and to install extra telephones there because at the time these decisions appeared correct; they were. Charles Meinhold of Brookhaven chose to encourage his old friend Margaret Reilly to ask for aid in radiation monitoring, making use of personal contacts to facilitate the response. Joe Deal chose to establish the five o'clock briefings at Capital City Airport to provide a badly needed exchange of data and information among different, and uncoordinated, federal and state agencies.

None of these choices was planned, but all were crucial. During the accident itself the decisions of plant operators frequently complicated an already difficult situation; in the Department of Energy response, the latitude for human choice and decision, on balance, was an advantage rather than a disadvantage. Things worked better because professional scientists made intelligent choices that no plan could have anticipated.

This is not to say that the Department of Energy was immune to the problems that plagued so many people during the hectic days of the accident. It too had difficulty obtaining accurate data from the crippled plant on the reactor's condition and the possibility of radiation releases and also had little evidence that radiological data passed along to other agencies was having any effect on the larger accident response. Internally, too, the Department of Energy experienced confusion as to the division of authority between Energy Technology and Environment, and between the command post and the Emergency Operations Center; the appropriate chain of command was not clearly understood throughout the Department. In the end the Department's response was most effective because key decisions were made locally by on-scene personnel, rather than in Washington by high officials.

Finally, it is significant that the quasi-military nature of the Department of Energy response to Three Mile Island proved the key to its efficiency. The response was directed by the Department's environmental branch. But military planning made the response to an unexpected civilian nuclear accident far better than it might otherwise have been. The Emergency Action Coordinating Team and the Emergency Operations Center were designed by military personnel to respond to a weapons-related accident and were organized under the Office of Military Application. The AMS/NEST teams likewise had more experience in dealing with downed satellites, airplanes or guided missiles than with nuclear power plant accidents. At Operation Morning Light in 1978 the Department learned about the virtues of a quasi-military response: get to the scene of battle as quickly as possible, put one person in charge of on-scene operations, bring more equipment than you think you will need, plan a longer stay than you first anticipate, avoid unnecessary contact with the press. Throughout the Three Mile Island accident the military nature of the Department of Energy response was an advantage — providing AMS/NEST with extra pilots and helicopters, transporting equipment by military aircraft, making evacuation facilities at U.S. Army barracks available, and so on. The only major criticism of military support at Three Mile Island was that more could have been used, especially in transporting people and equipment during an airlines strike and in coordinating logistics through the Joint Nuclear Accident Coordinating Center in Albuquerque, New Mexico.

The larger significance of the Department of Energy's response to the Three Mile Island accident is therefore paradoxical. In the public mind, the Department had inherited the mantle of the Atomic Energy Commission, a government agency created in wartime and geared to wartime needs: secrecy, contingency planning, centralized decision-making and isolation from public scrutiny. Yet these military qualities were precisely the virtues of emergency response during the battle of Three Mile Island. Comparing the accident to a military campaign is more than an exercise in metaphor. Unknown to the public, an army of scientists helped protect public health and safety during America's worst nuclear power accident. They did so in the face of unknown risks to their own safety. And the lack of public knowledge of that effort was itself no accident. For the most effective response turned out to be one which operated free of public scrutiny, but in the public interest.

AFTERWORD

Among the staff members whom the Department of Energy dispatched to Middletown, Pennsylvania, in response to the nuclear reactor accident on Three Mile Island, those from the Historian's Office arrived on the scene relatively late. More than a week had passed since the failure of the Unit 2 reactor when Chief Historian Richard G. Hewlett and his colleague, Jack M. Holl, reported to the Department of Energy command post at Capital City Airport. If the historians seemed tardy because they missed the height of the crisis, it was surprising, indeed unprecedented, that they were mobilized at all. Of the resources available to the Department of Energy during the emergency at Three Mile Island, the historians no doubt were among those who were the least professionally prepared to respond directly to the crisis. Virtually nothing in their graduate training or in their academic careers had prepared them to participate as professional historians in a major national emergency. In fact, they had not gone to Three Mile Island on their own initiative, but had been requested by Joe Deal and Herb Hahn, the principal Department of Energy respresentatives at the Capital City Airport command post.

Prior to Three Mile Island, the Department of Energy historians viewed their mission primarily in academic terms. Their primary task was to write book-length histories of their agency and its activities. Frequent reorganizations within the Department and its predecessor agencies, however, had somewhat altered their traditional role. Departing officials and defunct offices regularly deposited their papers in archives of the Historian's Office which had grown dramatically after the abolition of the Atomic Energy Commission. Department officials other than historians soon found these materials of current value. Herb Hahn, working on a project that required a knowledge of the past, discovered that the historian's skills and institutional memory could be useful in helping to solve practical, immediate problems. Subsequently, he proved perceptive in recognizing the historical importance of Three Mile Island.

Hahn, the first Department of Energy representative to arrive at Three Mile Island, was later joined by Joe Deal, who served as the senior Department of Energy representative. By the second day of the accident, Thursday, March 29, Hahn realized that he needed assistance in managing the flood of papers and notes which inundated the Command Post. In addition, both Deal and Hahn recognized the historical importance of the nation's first major commercial nuclear power accident, and the fact that the Department of Energy would have to prepare a major report of its activities. When he dropped exhausted into bed Thursday night, Hahn had already decided to seek assistance from the Historian's Office.

The great hydrogen bubble scare on Friday and Saturday, and President Carter's visit on Sunday, however, pushed thoughts about history firmly aside. Hahn was not able to catch his breath until Monday, April 2, when, with Deal's encouragement, he sent a request for the historians' help through regular department channels. Unfortunately, the historians were not told that they were wanted at the scene, but merely that they should contact the headquarters Emergency Operations Center in Germantown. They interpreted their directive in terms of verifying that the flow of information between headquarters and the field was being adequately documented for historical purposes.

Late Wednesday afternoon, exasperated that the historians had not yet arrived, Hahn called Holl personally. "Where are you guys?" Hahn asked. Surprised, Holl responded that the historians were indeed following Three Mile Island developments from headquarters. "Well, can you get up here and give us a hand?" Hahn demanded. After consulting with Hewlett, Holl promised that the historians would drive to Harrisburg the next morning. "By the way," Holl inquired, "what would you like us to do?" "Do?" Hahn responded incredulously, "How should I know. You're the professionals." Frequently critics of the foibles of bureaucracy, the historians were chastened by their own faltering response to the Three Mile Island crisis.

As they finally sped north towards Harrisburg, armed with note cards, tape recorders, and cameras, Hewlett and Holl speculated on how to fulfill their assignment. Certain tasks were obvious. They would observe, take notes, interview, collect data, and offer Deal and Hahn suggestions on how best to document the Department's activities. But as they drew nearer to Harrisburg, they also realized how uncertain they were about what their primary mission would be.

Fortuitously, Hewlett and Holl checked in at the Capital City Airport command post just as Robert Ferguson, Project Director, Office of Nuclear Energy Programs, arrived to review the Department's efforts at Three Mile Island. Along with Ferguson and his aides, the historians spent the day inspecting the Department's various operations. Starting at the command post, they visited

Trailer City, the Electric Power Research Institute "think tank" at the National Guard headquarters, and the press center in the Middletown gymnasium. Accompanied by Hap Lamonds, who served as their tour guide, they were briefed by government and industry officials, Joe Deal, Harold Denton, Milt Levinson, Herman Dieckamp, and William Lee. Remarkably, they also witnessed discussions on how to effect "cold shutdown" on the Unit 2. That night Hewlett and Holl shared a late dinner with Deal and Hahn during which they compared impressions of Three Mile Island and explored what the historians might do to offer continuing support.

Holl was awakened the following morning by Hewlett's impatient knocking on the motel room door. In the early hours, Hewlett had decided that the historians' most useful contribution would be their most traditional—to write a history of the Department's response to the Three Mile Island crisis. But there were problems. Hewlett and Holl were already writing full-time on a history of the Atomic Energy Commission. The rest of the professional staff was small and fully committed. A history of Three Mile Island would have to be written swiftly and well. The history should also be objective and professional. Over breakfast, Holl sketched a plan for hiring independent contractors to do the job, including estimated costs and time tables.

Deal quickly endorsed the idea and promised to assist in obtaining funds for the history project. From Deal's perspective, contractors could offer invaluable assistance collecting reliable documentation useful in forthcoming hearings while preparing an independent report of the Department's activities. While Deal obtained support for the plan from Ruth Clusen, Assistant Secretary for Environment, Hewlett and Holl returned to Germantown to work out the details.

The history was written by Philip Cantelon and Robert Williams while under contract to the Department of Energy. Holl served as project manager, but the authors determined the book's organization and analysis. For the Historian's Office, the project offered a major challenge, testing whether professional historians sponsored by the government could write a scholarly book which would be both useful to the department and credible to the public. The public will decide whether this hope has been realized.

Jack M. Holl
Chief Historian (Acting)

Bibliographical Essay

T
he published and unpublished documentation on the Three Mile Island accident of 1979 is immense. The Three Mile Island Records in the Historians' Office at the Department of Energy, Germantown, Maryland, constitute a major collection of this documentation. The Records include the Department of Energy's raw materials related to the accident (flight logs, Emergency Operations Center logs, analysis of samples from the national laboratory), various government reports and publications, newspaper and journal coverage, and various post-accident analyses. The comprehensive microfiche collection of pre- and post-accident documents assembled by the Nuclear Regulatory Commission is also part of the Records collection.

There are two major analyses of the accident that provide a useful overview and numerous policy recommendations. The *Report of the President's Commission on the Accident at Three Mile Island; The Need for Change: The Legacy of Three Mile Island* appeared in October 1979 and summarized the work of the President's Commission chaired by John Kemeny. *Three Mile Island: A Report to the Commissioners and the Public* (Washington, D.C., January 1980) is the report of the Nuclear Regulatory Commission's Special Inquiry Group, directed by Mitchell Rogovin; a second volume consisting of three separate parts of documentation and detailed studies appeared in May 1980. Neither of these reports pays particular attention to the Department of Energy monitoring and support effort. The most relevant unpublished materials from the Kemeny Commission are the "Report of the Office of Chief Counsel on Emergency Response," dated October 30, 1979, and the deposition before the Commission of L. Joe Deal, dated August 20, 1979.

In addition to these general and comprehensive studies, there are two important technical investigations. The first is the Nuclear Regulatory Commission's NUREG-0600, *Investigation into the March 28, 1979 Three Mile Island Accident by Office of Inspection and Enforcement* (Washington, D.C., August 1979), which details the technical course of the accident for the period March 28-March 30, 1979. The second is the May 10, 1979, report by a group of scientists assembled on an *ad hoc* basis by the Department of Health, Education, and Welfare entitled *Population Dose and Health Impact of the Accident at Three Mile Island Nuclear Station (A Preliminary Assessment for the Period*

March 28 through April 7, 1979). This latter report draws considerably upon radiological monitoring data collected by the Department of Energy.

The most useful materials of the Nuclear Regulatory Commission for our purposes were the series of bulletins distributed during the accident under the title "Preliminary Notification of Event or Unusual Occurrence" and the transcripts of the taped discussions of the Commissioners for March 30, March 31, and April 1, 1979, along with transcripts of the press conferences held by Commission Chairman Joseph Hendrie and Harold Denton.

The most useful Department of Energy material is to be found in the logs kept by various duty officers within the Department. The log of the Emergency Operations Center in Germantown is the most comprehensive, and includes a chronology of all incoming and outgoing messages and tape transcripts of telephone conversations. The log kept by the Brookhaven Radiological Assistance Team which responded to the accident is likewise of great value. Briefer chronologies were later compiled by the various national laboratories; the most germane, because of the work done on the hydrogen bubble problem, is the Idaho National Engineering Laboratory's "Chronology of Events at INEL related to the Three Mile Island Accident."

There is also an abundance of raw data collected by the various teams sampling for radiation on the ground and flight readings from the AMS/NEST helicopters. ARAC Instantaneous Air Concentrations, notebooks kept by Department personnel working out of Capital City Airport, and an extensive collection of oral interviews with scientists and Department of Energy officials comprise the remainder of the Records.

For background on the Departmental effort at Three Mile Island, the two most important reports are DOE/EV-0010, *Interagency Radiological Assistance Plan (IRAP)* (revised edition, March 1978) and the report by EG&G Aerial Measuring Systems, *An Aerial Radiological Survey of the Three Mile Island Station Nuclear Power Plant, August 1976* (Las Vegas, 1977). There is also a useful unpublished post-accident analysis of the Department's work entitled "Department of Energy Radiological Response to the Three Mile Island Accident," dated April 14, 1979.

FOOTNOTES
CHAPTER I

1. *Philadelphia Inquirer,* April 8, 1979, p. 1; Donelan reported "a whole bunch of steam coming out of the safety valve on top of Unit 2 — just clear steam" and said he "didn't think anything about it." The Garnishes described it as a "loud gush of steam." On Bill Whitlock, see *Pennsylvania Illustrated,* August 1979, p. 23.

2. United States Nuclear Regulatory Commission (USNRC), NUREG-0600, *Investigation into the March 28, 1979 Three Mile Island Accident by Office of Inspection and Enforcement,* August 1979, I-1-pp. 29-32. Hereafter cited as NUREG-0600.

   This final report revises several earlier accident chronologies:
   (a) USNRC, IE (Office of Inspection and Enforcement), "Chronology of Events: TMI-2 Incident," April 5, 1979;
   (b) USNRC, "TMI-2 Interim Operational Sequence of Events as of May 8, 1979," USNRC Briefing;
   (c) USNRC, EI, "TMI-2 Radiological Sequence of Events as of June 19, 1979," June 1979.

   For an account of the accident from the perspective of Babcock and Wilcox, the prime contractor for Unit 2, see *Nuclear News,* XXII, no. 9 (July 1979), pp. 30-31. According to B&W officials, the accident was due to six factors: 1) auxiliary feedwater valves were closed, not open, when the main feedwater was lost to the system; 2) the pressurizer relief valve stuck open for over two hours; 3) there had been an "inappropriate emphasis" on the pressurizer water level indicator, without considering the actual pressure in the pressurizer; 4) the emergency core coolant system high-pressure injection pumps were prematurely shut down; 5) primary coolant pumps were shut down; 6) containment sump pumps were operating during high-pressure injection, discharging radioactive water to the auxiliary building.

3. NUREG-0600, I-1 pp. 34-35; see also the transcripts of the "Meeting of the President's Commission on the Accident at

Three Mile Island," May 30, 1979, pp. 153-154. Hereafter cited as President's Commission, "Meeting."

4. NUREG-0600, p. 2; President's Commission, "Meeting," May 30, 1979, pp. 178-180.

5. *St. Louis Post-Dispatch*, July 25, 1979, p. 3B.

6. *Pennsylvania Illustrated*, August 1979, p. 25.

7. According to subsequent testimony before the President's Commission, surveillance tests of the two valves were performed on March 26, 1979, over a three-hour period. Those involved maintained that the valves were then re-opened, although the written checkoff sheets that could have verified that fact were "thrown in the trash can." Operators did testify that the valves could have been closed at the valves themselves, at extension controls inside the plant, or in the control room. Gary Miller claimed that "they could have been closed possibly by air from a local switch or from the control room by air." See President's Commission, "Meeting," May 30, 1979, pp. 78-80, and May 31, 1979, p. 41.

8. NUREG-0600, Appendix I-A, p. 19; President's Commission, "Meeting," May 30, 1979, p. 152.

9. NUREG-0600, Appendix I-A, p. 27; President's Commission, "Meeting," May 30, 1979, p. 168.

10. NUREG-0600, p. 14 and II-F pp. 5-6.

11. NUREG-0600, p. 15.

12. President's Commission, "Meeting," May 30, 1979, p. 169.

13. President's Commission, "Meeting," May 31, 1979, p. 131.

14. *Ibid.*, p. 64.

15. Gary P. MIller, "TMI Station. March 28, 1979 Incident. Statement by G. P. Miller, Station Manager," May 7, 1979, p. 7; President's Commission, "Meeting," May 31, 1979, pp. 5-6.

16. NUREG-0600, Appendix I-A, pp. 38-39.

17. NUREG-0600, p. 5 and Appendix I-A, p. 58.

18. NUREG-0600, Appendix II-A, pp. 5-8.

19. NUREG-0600, Appendix II-A, pp. 8-9. *St. Louis Post-Dispatch,* July 26, 1979, p. 3B. President's Commission, "Meeting," May 31, 1979, pp. 68-71.

20. *Pennsylvania Illustrated,* August 1979, p. 23.

CHAPTER II

1. Metropolitan Edison Company, "Environmental Report, 1971," December 31, 1971 in the Nuclear Regulatory Commission's microfiche collection, *Pre-Incident Documents*, 7904300413, pp. 17, 67, 73, 91-92.

2. *Ibid.*, pp. 91-93; extract from Metropolitan Edison, *Three Mile Island Unit 2 Final Safety Analysis Report*, hereafter cited as *Final Safety Analysis Report*; Nuclear Regulatory Commission Document Number 50420, Section C, "Safety-Related Design Criteria," p. 6, in "Three Mile Island Nuclear Plant Accident Report for the Assistant Secretary of Energy Technology," April 1, 1979, Three Mile Island Records, Historian's Office, Department of Energy, Germantown, Maryland; collection hereafter cited as Three Mile Island Records.

3. Metropolitan Edison, "Environmental Report, 1971," Nuclear Regulatory Commission, *Pre-Incident Documents*, 7904300413, pp. 89-90.

4. *Baltimore Sun*, April 8, 1979; Metropolitan Edison, "Environmental Report, 1971," Nuclear Regulatory Commission, *Pre-Incident Documents*, 7904300413, pp. 148, 92-93.

5. *Baltimore Sun*, April 8, 1979.

6. *Ibid.*, April 8, 1979.

7. *Ibid.*, April 8, 1979; Metropolitan Edison, "Environmental Report, 1979," Nuclear Regulatory Commission, *Pre-Incident Documents*, 7904300413, pp. 17, 23, 25, 321, 67.

8. NUREG-0600, pp. v-vi.

9. *Final Safety Anaylsis Report*, pp. 9.2.3.; 15.1.8; 15.1.17.

10. NUREG-0600, pp. I-4-2; I-1-39 and 44; I-1-52.

11. NUREG-0600, pp. I-1-45; *Engineering News-Report*, April 5, 1979, p. 13.

12. NUREG-0600, pp. I-1-54; I-4-46; President's Commission, May 30, 1979, pp. 104-110.

13. W. J. Lanouette, "No Longer can the NRC say...," *Bulletin of the Atomic Scientists*, Vol. 35, No. 6 (June 1979), pp. 7-8; *New York Times*, July 19, 1979, p. D8 and July 20, 1979, p. A1; T. M. Novak, memorandum, January 10, 1978, "Loop Seals in Pressurizer Surge Line." The Babcock & Wilcox memoranda were dated November 1, 1978 (Kelly) and February 9, 1978 (Dunn), copies in Three Mile Island Records.

14. *Final Safety Analysis Report*, Appendix 13A, "Three Mile Island Site Emergency Plan."

15. NUREG-0600, pp. II-2, 8-21, contained the Metropolitan Edison plan and its actual operation during the accident.

16. *Ibid.*, Appendix 13A, 4.3.2.

17. Department of Environmental Resources, Bureau of Radiological Health, "Plan for Nuclear Power Generating Station Incidents," September 1977, copy in Three Mile Island Records.

18. *Ibid.*, p. III-4.

19. *Ibid.*, p. III-7.

20. *Ibid.*, pp. IV-2, IX-4. Brookhaven National Laboratory did not have the aerial monitoring capability itself, but could call for assistance on the Department of Energy's Aerial Measuring Systems/Nuclear Emergency Search Team resources at Andrews Air Force Base and Las Vegas, Nevada.

21. Bureau of Radiological Health, *Three Mile Island Nuclear Station*, "Annex to the Pennyslvania Plan for the Implementation of Protective Action Guides," Parts V, VI, VII, VII-17, copy in Three Mile Island Records.

22. *Final Safety Analysis Report*, Appendix 13A, 3.5.1. and 3.5.3.

23. Author's telephone interview with A. E. Fritzsche, October 4, 1979, Three Mile Island Records.

24. Author's telephone conversation with John F. Doyle, October 24, 1979; interview with L. Joe Deal, October 16, 1979, Three Mile Island Records. Doyle manages the Aeri-

al Measurement Operations for EG&G, Inc., the present contractor of the Aerial Measuring Systems for the Department of Energy. The Windscale accident also led to the first decision within the Atomic Energy Commission on a national plan for radiological assistance, Atomic Energy Commission, Commissioner's Meeting 1334, February 12, 1958.

25. Interview with John F. Doyle, October 30, 1979, Three Mile Island Records; interview with L. Joe Deal, October 16, *ibid.*; author's telephone conversation with John F. Doyle, October 24, 1979. While concentrating on government and commercial nuclear facilities, the Aerial Radiation Measuring Systems group also helped search for the lost nuclear submarine *U.S.S. Thresher,* on the chance that some radiation had escaped from the sunken vessel. No radioactivity was sighted; *ibid.* By 1976, the measuring system had changed its name from Aerial Radiation Measuring Systems to Aerial Measuring Systems and had been incorporated into a program which also responded to acts of nuclear terrorism called Nuclear Emergency Search Team. Subsequently, the aerial monitoring teams have used the combined acronym AMS/NEST for identification.

26. Author's telephone interview with A. E. Fritzsche, October 4, 1979, *ibid.; Final Safety Analysis Report,* Vol. 1, 2.2-1-22; EG&G, "An Aerial Radiological Survey of the Three Mile Island Station, August, 1976" (Las Vegas, 1977), p. 2; *ibid.*, pp. iii, 11. In November 1976, EG&G resurveyed the Harrisburg area to measure fallout levels from the Chinese weapons test and found an increase in the level of radioactivity. Interview with John F. Doyle, October 30, 1979, Three Mile Island Records.

27. In the late 1950s and early 1960s thirty-one sites had radiological assistance teams. Author's telephone interview with F. Raymond Zintz, November 8, 1979, *ibid.*; interview with L. Joe Deal, October 16, 1979, *ibid.*

28. Interview with Charles Meinhold, Brookhaven National Laboratory, August 14, 1979, *ibid.*; information on the instrument is in C. Distenfeld and J. Klemish, "An Air Sampling System for Evaluating the Thyroid Commitment Due

to Fission Products Released from Reactor Containment," NUREG/CR-0314, December 1978, copy in Three Mile Island Records.

29. Energy Research and Development Administration Manual 0601, Annex 15, "Atmospheric Release Advisory Capability," July 1977, p. 2; author's telephone conversation with Marvin Dickerson of the Lawrence Livermore Laboratory, November 14, 1979, Three Mile Island Records.

30. See F. Raymond Zintz, "The United States Atomic Energy Commission Radiological Assistance Program and Allied Activities," speech delivered at The Hague, Netherlands, December 15, 1961, copy in Three Mile Island Records. Zintz was the "father" of the radiological assistance program.

31. *Ibid.*, p. 4.

32. United States Department of Energy, "Interagency Radiological Assistance Plan, March 1978," DOE/EV-0010, p. 8, copy in Three Mile Island Records.

33. "Agreement between the United States Energy Research and Development Administration and the Nuclear Regulatory Commission for Planning, Preparedness, and Response to Emergencies," March 8, 1977, in Three Mile Island Records; EG&G Aerial Measuring Systems, "An Aerial Radiological Survey of the Three Mile Island Station Nuclear Power Plant, August, 1976," (Las Vegas, 1977), p. 1,; see "Interagency Radiological Assistance Plan, 1978," pp. 27, 24.

34. Author's telephone conversation with Gerald Combs, November 14, 1979, Three Mile Island Records. While manual chapters carried the weight of law, those requirements did not carry over when the Energy Research and Development Administration became the Department of Energy, according to Combs. Nonetheless, the manual chapter had not been superseded by any new guidelines and two of the chapter's authors, J. Beaufait and Herbert Hahn, are still active in the Department's emergency response program. Beaufait works for the Environmental Protection and Public Safety Branch in the Office of En-

vironmental Compliance and Overview and is deeply involved in the activities of the Emergency Operations Center. Hahn is the Department's site representative to the Aerial Measurement Systems/Nuclear Emergency Search Team operation at Andrews Air Force Base in Maryland.

35. Energy Research and Development Administration, Manual 0601, "Emergency Planning, Preparedness, & Response Program," Annex 1, pp. 11, 3.

36. *Ibid.*, Annex 7, pp. 10, 11-13.

37. *Ibid.*, Annex 7, pp. 15, 25-29.

38. General Accounting Office, Report EMD-78-110, March 30, 1979, p. i, copy in Three Mile Island Records.

39. S. McCracken, "The Harrisburg Syndrome," *Commentary*, Vol. 67, No. 6 (June 1979), pp. 27-39 effectively summarizes the general background to the Three Miles Island accident in terms of the politics of nuclear power.

CHAPTER III

1. Interview with Nathaniel Greenhouse, August 21, 1979, Three Mile Island Records.

2. Interview with Nathaniel Greenhouse, *ibid.*; Brookhaven Communications Log, Three Mile Island Nuclear Power Plant Incident, March 28, 1979, Three Mile Island Records; hereafter cited as Brookhaven Log.

3. Interview with Nathaniel Greenhouse, August 21, 1979, Three Mile Island Records.

4. Brookhaven Log, March 28, 1979 *ibid.*; interview with Ed Lessard, August 14, 1979, *ibid.*; interview with Robert Casey, August 14, 1979, *ibid.*; interview with Robert Friess, August 13, 1979, *ibid.*

5. Brookhaven Log, March 28, 1979, *ibid*; interview with David Schweller, August 13, 1979, *ibid*; author's telephone interview with David Schweller, December 6, 1979.

6. Brookhaven Log, March 28, 1979, Three Mile Island Records; see also Robert Bores, "State, Federal, & Misc. Contacts," March 28, 1979, copy in Three Mile Island Records, hereafter cited as Bores Log; interview with David Schweller, August 13, 1979, *ibid.*

7. Interview with Charles Meinhold, August 14, 1979, *ibid.*; Thomas M. Gerusky to Bruce T. Lundin, June 18, 1979, copy in Three Mile Island Records; interview with Robert Friess and David Schweller, August 13, 1979, *ibid.*

8. *Ibid.*; Report of the President's Commission on the Accident at Three Mile Island, *The Need for Change: The Legacy of TMI* (Washington, D.C., 1979), pp. 101-103; hereafter cited as President's Commission, *Report.*

9. *Ibid.*, p. 101.

10. *Ibid.*, p. 101; Department of Environmental Resources, Bureau of Radiological Health, "Plan for Nuclear Power Generating Station Incidents," September 1977, II/3-4.

11. President's Commission, *Report*, pp. 101-103.

12. *Ibid.*, pp. 102-103; Gerusky to Lundin, June 18, 1979, Three Mile Island Records.

13. President's Commission, *Report*, pp. 104, 106.

14. *Ibid.*, p. 106.

15. Gerusky to Lundin, June 18, 1979, Three Mile Island Records; Bores Log, March 28, 1979, *ibid.*; interview with Robert Friess, August 13, 1979, *ibid.* Because the field analyses on the iodine 131 levels were only estimates, or better, "guesstimates," state police flew the samples to the helipad at Harrisburg's Holy Spirit Hospital and they were driven to the state lab. The laboratory's spectrum analysis revealed no radioactive iodine. Nonetheless, decisions were made based on the misinformation. See Bores Log, March 28, 1979, *ibid.* and Gerusky to Lundin, June 18, 1979, *ibid.*

16. Bores Log, March 28, 1979, *ibid.*; interview with David Schweller and Robert Friess, August 13, 1979, *ibid.*; Bores Log, March 28, 1979, *ibid.*; interview with Charles Meinhold, August 14, 1979, *ibid.*

17. Emergency Operations Center Chronology, "Three Mile Island Incident," March 28, 1979; hereafter cited as Emergency Operations Center Log, Three Mile Island Records; the initial alerts for Bettis and Oak Ridge were informal, see Operation Ivory Purpose File, Three Mile Island Records.

18. President's Commission, *Report*, 104, 106; Emergency Operations Center Log, March 28, 1979, Three Mile Island Records. The Kemeny Commission *Report* is in error on the important point of which agency called in which Department of Energy resource for assistance at Three Mile Island: "... both Pennsylvania's Bureau of Radiation Protection and the Nuclear Regulatory Commission requested the Department of Energy to send a team from Brookhaven National Laboratory to assist in monitoring environmental radiation," states the *Report* on page 106. The Nuclear Regulatory Commission requested only the AMS/NEST aerial survey. The state requested Brookhaven's radiological assistance teams.

19. Interview with Roy E. Lounsbury, September 9, 1979, Three Mile Island Records.

20. Interview with Herbert Feinroth, September 10, 1979, *ibid.*

21. Interview with L. Joe Deal, October 17, 1979, *ibid.*

22. L. J. Beaufait, memorandum, March 28, 1979, *ibid.* At the time Beaufait was Deal's assistant and made these notes at the meeting.

23. *Ibid.;* Paul Gudiksen, taped memorandum, April 1979, Three Mile Island Records. Gudiksen, a scientist from the Livermore Laboratory, worked on the development of the ARAC system and came to Three Mile Island to work at the ARAC center at the Department of Energy command post.

24. Interview with G. Robert Shipman, July 18, 1979, *ibid.*

25. *Ibid.*

26. Interview with Herbert F. Hahn, August 21, 1979, *ibid.;* interview with Captain C. Creel and Lieutenant R. Bauland, July 27, 1979, *ibid.; Capitol Flyer,* May 25, 1979, in *ibid.*

27. Interview with Colonel Roger J. Leushow, July 27, 1979, ibid.; Leushow was Wells' replacement at the 1st Helicopter Squadron; Wells quoted in *Capitol Flyer,* May 25, 1979; ten men volunteered from the 1st Helicopter Squadron for Three Mile Island duty, *ibid.*

28. Emergency Operations Center Log, March 28, 1979, *ibid.;* Bores Log, March 28, 1979, *ibid.*

29. Interview with Herbert F. Hahn, August 21, 1979, *ibid.;* Emergency Operations Center Log, March 28, 1979, *ibid.;* interview with Herbert F. Hahn, August 21, 1979, *ibid.*

30. The fact that there were two airports serving Harrisburg created some confusion. When the AMS/NEST party arrived they ordered two rental cars. But the automobiles never came; they went instead to the Harrisburg International Airport. The mistake was not repeated. Interview with G. Robert Shipman, July 18, 1979, *ibid.*

31. Emergency Operation Center Log, March 28, 1979, *ibid.;* interview with Herbert F. Hahn, August 21, 1979, *ibid.*

32. Interview with Roger Leushow, July 27, 1979, *ibid.*

33. Interview with Nathaniel Greenhouse, August 21, 1979, *ibid.;* interview with Roger Leushow, August 27, 1979, *ibid.* See also Brookhaven Log, March 28, 1979, *ibid.*

34. Interview with David Schweller, August 13, 1979, *ibid.;* interview with Edward Lessard, August 13, 1979, *ibid.;* Brookhaven Log, March 28, 1979, *ibid.*

35. Interview with David Schweller, August 13, 1979, *ibid.;* interview with Nathaniel Greenhouse, August 21, 1979, *ibid.;* interview with Edward Lessard, August 13, 1979, *ibid.;* Brookhaven Log, March 28, 1979, *ibid.*

36. Interview with G. Robert Shipman, July 18, 1979, *ibid.*

37. *Ibid.*

38. *Ibid.;* interview with Jac Watson, July 17, 1979, *ibid.;* Bores Log, March 28, 1979, *ibid.*

39. Interview with Herbert F. Hahn, August 21, 1979, *ibid.*

40. *Ibid.*

41. Interview with David E. Patterson, September 11, 1979, and Ed Patterson Memorandum, March 28, 1979, *ibid.;* Brookhaven Log, March 28, 1979, *ibid.;* Emergency Operations Center Log, March 28, 1979, *ibid.*

42. Interview with Edward Lessard, August 13, 1979, *ibid.;* interview with William R. Casey, August 14, 1979, *ibid.;* interview with Robert Friess, August 13, 1979, *ibid.*

43. *Ibid.*

44. Interview with David E. Patterson, September 11, 1979, *ibid.;* interview with G. Robert Shipman, July 18, 1979, *ibid.;* interview with Herbert F. Hahn, August 21, 1979, *ibid.;* Brookhaven Log, March 28, 1979, *ibid.*

45. Interview with David Schweller and Robert Friess, August 13, 1979, *ibid.;* Bores Log, March 28, 1979, *ibid.*

46. Cronkite quoted in *TV Guide,* August 4, 1979, p. 6.

47. Interview with Herbert F. Hahn, August 21, 1979, Three Mile Island Records.

48. Interview with Robert Friess, August 13, 1979, *ibid.;* interview with David E. Patterson, September 11, 1979, *ibid.;* Brookhaven Log, March 29, 1979, *ibid.;* AMS/NEST Flight Log, March 29, 1979, *ibid.*

49. Paul Gudiksen, memorandum, no date, *ibid.*

50. Interview with Roy E. Lounsbury, September 7, 1979, *ibid.;* Emergency Operations Center Log, March 29, 1979, *ibid.;* Oak Ridge Memorandum for Operation Ivory Purpose, no date, Ivory Purpose File, *ibid.;* interview with L. Joe Deal, October 17, 1979, *ibid.*

51. Interview with G. Robert Shipman, July 18, 1979, *ibid.;* AMS/NEST Flight Log, March 29, 1979, *ibid.;* Emergency Operations Center Log, March 29, 1979, *ibid.*

52. The Nuclear Regulatory Commission press release of March 29, 1979, is in the Emergency Operations Center Log, Tab. 8, *ibid.;* see also the *Washington Star,* March 29, 1979.

53. Interview with Jac Watson, July 17, 1979, *ibid.;* President's Commission, *Report,* 114. Alternatively, if the Nuclear Regulatory Commission had not known of the 3,000 milli-rem release and it had been uncontrolled rather than plan-ned, then there was no reason for the Nuclear Regulatory Commission to notify the Capital City command post. In any event, the release created "no great concern" at the Commission. Still, the fact that the Metropolitan Edison helicopter was hovering above the vent when the release took place indicates that the company had planned the venting. See President's Commission, *Report,* 114 and NUREG-0600, II-A-58.

54. Interview with Robert Friess, August 13, 1979, *ibid.;* interview with David E. Patterson, September 11, 1979, *ibid.;* interview with G. Robert Shipman, July 18, 1979, *ibid.*

55. Interview with L. Joe Deal, October 17, 1979, *ibid.;* in his deposition to the President's Commission, Deal placed his visit to the Nuclear Regulatory Commission trailers on Fri-day morning; see Deposition of L. Joe Deal, August 20, 1979, a copy of which is in the Three Mile Island Records. On flying conditions at Three Mile Island see interview with Jac Watson, July 27, 1979, *ibid.*

142

56. Interview with Herbert F. Hahn, August 21, 1979, *ibid.; see* also Hahn's interview with Chief Donald Grow, April 10, 1979, *ibid.* Grow was chief of the Capital City Airport police.

57. Interview with L. Joe Deal, October 17, 1979, *ibid.;* interview with Roy E. Lounsbury, September 7, 1979, *ibid.* This policy of keeping a low profile remained in effect throughout the accident. Dale Meyers, the Deputy Secretary, Ruth C. Clusen, Assistant Secretary for Environment, and Assistant Secretary for Defense Programs Duane C. Sewell planned to tour the Capital City command post on April 4. The previous day Colonel Roy Lounsbury had cleared an Andrews Air Force Base helicopter to pick up the three Department of Energy officials at the Pentagon and ferry them to Pennsylvania. But on the morning of the fourth the trip was scrubbed. Apparently, according to Lounsbury, the cancellation came from the Secretary's office. See Emergency Operations Center Log, April 3, 1979 and April 4, 1979, *ibid.;* interview with Roy E. Lounsbury, September 7, 1979, *ibid.*

58. "Challenging the Atomic Energy Commission on Nuclear Reactor Safety: The Union of Concerned Scientists," in Joel Primack and Frank von Hippel, *Advise and Dissent* (New York, 1974), 208-232.

59. *TV Guide,* August 4, 1979, pp. 6, 8; Governor Richard Thornburgh, Press Conference, printed in J. C. Staley and R. R. Seip, *Three Mile Island: A Time of Fear* (Harrisburg, 1979).

60. President's Commission, *Report,* 115; Emergency Operations Center Log, March 30, 1979, Three Mile Island Records; President's Commission, *Report,* 116.

CHAPTER IV

1. Cantril, Hadley, *The Invasion from Mars* (Princeton, 1940).

2. On the Friday morning releases from the plant, see NUREG-0600, IIA, pp. 66-67 and II-2, p. 23. The best short summary of the day's events is in the President's Commission, *Reports*, pp. 116-125.

3. Floyd's testimony on his activities on Friday is in the President's Commission, Transcript, May 31, 1979, especially pp. 172-173 and 178-179. See also J. Graham, "President's Commission Confronts the 'Truth'," *Nuclear News*, XXII, no. 9 (July 1979), pp. 20-30; and NUREG-0600, II-2, p. 23.

4. President's Commission, Transcript, August 2, 1979, pp. 9-10 and 29-30 on the testimony of Kevin J. Molloy.

5. *Ibid.*, p. 307, contains the testimony of Harold E. Collins. For a summary of other testimony on the Friday morning telephone calls, see John Graham, "Recollections of a Day in March," *Nuclear News*, XXII, no. 11 (September 1979), pp. 30-32.

6. Nuclear Regulatory Commission, Meeting, March 30, 1979, Transcript, p. 8.

7. President's Commission, Transcript, August 2, 1979, pp. 41-42 (Oran K. Henderson) and pp. 311-312 (Harold Collins). Harold Denton testified on May 31, 1979, President's Commission, Transcript, p. 304, that "we, in effect, directed Mr. Collins to call the state, so he did not do it on his own initiative." The President's Commission later concluded that Collins was "apparently selecting the distance on his own." See President's Commission, *Report*, p. 118.

8. President's Commission, Transcript, August 2, 1979, testimony of William Scranton, p. 222; testimony of Thomas Gerusky, p. 96.

9. President's Commission, Transcript, May 31, 1979, testimony of Charles O. Gallina, pp. 257 ff.; Nuclear Regulatory Commission, Closed Commission Meeting, March 30, 1979, Transcript, pp. 24-27; *Washington Post*, April 8, 1979, p. A18; President's Commission, *Report*, p. 119.

10. Nuclear Regulatory Commission, Closed Commission Meeting, March 30, 1979, Transcript, pp. 14, 19; President's Commission, *Report,* p. 119.

11. Nuclear Regulatory Commission, Closed Commission Meeting, March 30, 1979, p. 55; around eleven Joseph Hendrie told Denton at Bethesda that "the President just called over, I think you had better go down to the site. He'd like to see a senior officer and I think you are it."

12. The full transcript of the Thornburgh press conference is in Staley and Seip, *Three Mile Island: A Time of Fear.*

13. U.S. Nuclear Regulatory Commission, Preliminary Notification-79-67B (March 30, 1979), p. 1. The preliminary notification also provided specific pressure and temperature readings for the period from ten a.m. Thursday to five a.m. Friday morning, and noted "significant releases of iodine and noble gases from the fuel." Low off-site radiation readings were reported. The notification also observed that the plant had released about fifty thousand gallons of contaminated wastewater on Thursday to the Susquehanna River and had released more on Friday with Nuclear Regulatory Commission permission.

14. *Ibid.,* p. 4.

15. Nuclear Regulatory Commission, Closed Commission Meeting, March 30, 1979, Transcript, p. 61.

16. *Ibid.,* p. 68.

17. *Ibid.,* pp. 72, 77-78, 83-84.

18. *Washington Post,* April 8, 1979, p. A18 and April 9, 1979, p. A1.

19. *Washington Post,* April 10, 1979, p. A18.

20. Jack Crawford to John Deutch, March 30, 1979, "Report of White House Meeting on Three Mile Island Nuclear Plant Situation," Energy Technology Folder, Three Mile Island Records.

21. Nuclear Regulatory Commission, Closed Commission Meeting, March 30, 1979, Transcript, pp. 106-107, 108, 110.

22. *Ibid.*, pp. 127-128; Preliminary Notification-79-67C; copy in Emergency Operations Center Log, Three Mile Island Records.

23. UPI Press Release, March 30, 1979, copy, *ibid.* Shortly after the UPI story broke, Hendrie said that "Dudley Thompson got wandering off about what might happen if the gas bubble expanded," Nuclear Regulatory Commission, Closed Commission Meeting, March 30, 1979, Transcripts, p. 156, Nuclear Regulatory Commission Folder, *ibid.*

24. *Harrisburg Patriot,* March 31, 1979, pp. 1, 5, 18; *New York Times,* March 31, 1979, pp. Al, A19; *Washington Post,* March 31, 1979, p. A2.

25. Nuclear Regulatory Commission, Closed Commission Meeting, March 30, 1979, Transcript, pp. 207-209, Nuclear Regulatory Commission Folder, Three Mile Island Records.

26. President's Commission, Transcript, August 2, 1979, pp. 74-75, President's Commission File, *ibid.*

27. *New York Times,* April 16, 1979.

28. *T.V. Guide,* August 4, 1979, pp. 8, 11.

29. Interview with Nathaniel Greenhouse, August 21, 1979, Three Mile Island Records; interview with William R. Casey, August 14, 1979, *ibid.*; interview with David Schweller and Robert Friess, August 13, 1979, *ibid.*; interview with William R. Casey, August 14, 1979, *ibid.*

30. Interview with Nathaniel Greenhouse, August 21, 1979, *ibid.*

31. Interview with William R. Casey, August 14, 1979, *ibid.*; interview with Charles Meinhold, August 14, 1979, *ibid.*

32. Brookhaven Log, March 30, 1979, *ibid.*

33. Interview with Walter Frankhauser, July 27, 1979, *ibid.*

34. Interview with Jac Watson, July 17, 1979, *ibid.*; interview with Captain C. Creel and Lieutenant R. Bauland, July 27, 1979, *ibid.*

35. Joe Deal to W. J. McCool, transcript of telephone conversation, March 30, 1979, Emergency Operations Center Log, Tab. 10, *ibid.*

36. Emergency Operations Center Log, Tab. 11, *ibid.*

37. Emergency Operations Center Log, Tab. 12, *ibid.*

38. Emergency Operations Center Log, Chronology, Tab. 13, p. 7, *ibid.*

39. "Chronology of Events at Idaho National Engineering Laboratory related to Three Mile Island," pp. 1-3, Idaho National Engineering Laboratory File, Three Mile Island Records; *Washington Post,* April 18, 1979, p. 16. The hydrogen bubble problem had been anticipated by the Advisory Committee on Reactor Safeguards in 1969; in 1971 the Committee warned of possible hydrogen build-up after a loss of coolant accident and recommended that power plant license applicants submit "a proposed design for hydrogen control, including provisions for inerting." See *Science,* Vol. 204 (May 25, 1979), pp. 794-795 for references to the Advisory Committee on Reactor Safeguards reports.

40. This account of the White House meeting is based on an interview with Herbert Feinroth, September 10, 1979, Three Mile Island Records.

41. Emergency Operations Center Log, Chronology, p. 8, *ibid.*

42. Emergency Operation Center Log, Tab 14, *ibid.*

43. *Ibid.*

44. *Ibid.*

45. David E. Patterson, "Minutes of an Environmental Monitoring Coordination Meeting," March 30, 1979, transcript of tape, *ibid.*

46. *Ibid.*

47. President's Commission, Transcript, May 31, 1979, p. 353; on Friday morning Denton said that "we did hear from the Aerial Radiation Measuring Systems people that they were averaging something on the order of an R an hour at 600 feet above the plant," when the Aerial Measuring Systems plume flight had not been flown that morning. Villforth's statement was given to the President's Commission, Transcript, August 2, 1979, p. 243.

48. S. D. Brunn, J. H. Johnson, D. J. Ziegler, *Final Report on a Social Survey of Three Mile Island Area Residents*, (East Lansing, Michigan, August 1979), p. 41.

49. *Ibid.*, pp. 39, 47, 52, 152. Oran K. Henderson mentioned the Chinese rumor in his testimony before the President's Commission, Transcript, August 2, 1979, p. 80, President's Commission Folder, Three Mile Island Records.

CHAPTER V

1. *Washington Post*, April 10, 1979, p. 5.

2. Interview with Herbert Feinroth, September 10, 1979, Three Mile Island Records.

3. President's Commission, *Report*, pp. 123, 133; Nuclear Regulatory Commission, Preliminary Notification-79-67F (March 31, 1979) and Preliminary Notification-79-67G (April 1, 1979), copies in Emergency Operations Center Log, Three Mile Island Records.

4. *Washington Post*, April 9, 1979, p. A15; President's Commission, *Report*, p. 130.

5. *New York Times*, April 1, 1979, p. 31; *Harrisburg Patriot*, April 1, 1979, p. 2.

6. "Chronology of Events at HEW regarding Three Mile Island," May 17, 1979, copy in Health, Education and Welfare Folder, Three Mile Island Records; *Harrisburg Patriot*, April 2, 1979, p. 30. Iodine-131 found in milk samples measured thirty-six picocuries per liter, compared with the Environmental Protection Agency protective action level of 12,000 picocuries per liter; cesium-137 was found at forty-six picocuries per liter, compared with a protective action level of 340,000 picocuries per liter.

7. President's Commission, *Report*, p. 131; *Washington Post*, April 10, 1979, p. 18. There were 24,527 people living within five miles of Three Mile Island, 133,672 within ten miles, and 636,073 within twenty miles.

8. "Chronology of Events at HEW regarding Three Mile Island," May 17, 1979, p. 4, Health, Education and Welfare File, Three Mile Island Records; "Report of the Office of the Chief Counsel on Emergency Response," October 30, 1979, pp. 93, 95, President's Commission File, *ibid.*

9. Nuclear Regulatory Commission, Closed Commission Meeting, March 31, 1979, Transcript, pp. 46, 52, 59, 63, Nuclear Regulatory Commission File, *ibid.*; transcript of Hendrie press conference, pp. 3, 5, 13, *ibid.*

10. Nuclear Regulatory Commission, Closed Commission Meeting, *ibid.*; President's Commission, *Report*, p. 129.

11. *New York Times,* April 2, 1979, pp. A4, A12; *Washington Star,* April, 2, 1979, 48.

12. "Report of the Office of Chief Counsel on Emergency Response," October 30, 1979, pp. 97, 108, 149, President's Commission File, Three Mile Island Records.

13. *Washington Post,* April 10, 1979, p. 15.

14. "Report of the Office of Chief Counsel on Emergency Response," October 30, 1979, p. 125, President's Commission File, Three Mile Island Records.

15. *Ibid.,* p. 126.

16. *Washington Post,* April 10, 1979, p. 15; *New York Times,* April 2, 1979, p. A1.

17. *Washington Post,* April 10, 1979.

18. President's Commission, *Report,* p. 134; Nuclear Regulatory Commission, Closed Commission Meeting, April 1, 1979, pp. 97, 118, Nuclear Regulatory Commission File, Three Mile Island records.

19. AMS/NEST Flight Log, March 31, 1979, Three Mile Island Records.

20. Interview with G. Robert Shipman, July 18, 1979, *ibid,;* interview with Thomas Maguire, July 18, 1979, *ibid.;* interview with Captain C. Creel and Lieutenant R. Bauland, July 27, 1979, *ibid.*

21. Interview with William R. Casey, August 14, 1979, *ibid.*

22. Emergency Operations Center Log, March 31, 1979, Tab. 17, *ibid.*

23. Emergency Operations Center Log, Chronology, p. 10, *ibid.*

24. Brookhaven Log, pp. 10-12, *ibid.;* Emergency Operations Center Log, Chronology, p. 16.

25. Emergency Operations Center Log, Chronology, p. 10, *ibid.*

26. Interview with Roy E. Lounsbury, September 7, 1979, *ibid.*

27. President's Commission, Testimony of L. Joe Deal, August 20, 1979, pp. 41-42, President's Commission File, *ibid.*

28. Emergency Operations Center Log, March 31, 1979, Tab 18, *ibid.*

29. Emergency Operations Center Log, March 31, 1979, Tab 19, *ibid.*

30. Emergency Operations Center Log, March 31, 1979, Tabs 22-24, *ibid.*

31. David E. Patterson, "Recollections," transcribed tape, *ibid;* President's Commission, Testimony of L. Joe Deal, August 20, 1979, p. 39, *ibid.* The Bettis and and Brookhaven teams were already in place, although some Brookhaven personnel who had gone home on Thursday returned to Three Mile Island on Saturday. The four-man Argonne team led by Ed Jacewsky, arrived Friday evening around eight, followed Saturday morning at six a.m. by a second five-man team in a van. The Oak Ridge team, consisting of a Department representative and five health physicists, arrived at nine Friday evening and reported to Trailer City at nine-thirty Saturday morning to assist the Nuclear Regulatory Commission. On Saturday, Paul Gudikson arrived from Livermore Laboratory to assist Dickerson with ARAC operations and set up an open telephone line to California. In addition, a six-man Mound Laboratory team from Dayton, Ohio, arrived at eleven a.m. Saturday. On Monday they helped the Food and Drug Administration place dosimeters around the area. See "DOE Radiological Response to the Three Mile Island Accident," April 14, 1979, pp. C5-7 and D2, Three Mile Island Records.

32. Emergency Operations Center Log, March 31, 1979, Tab. 24, *ibid.*

33. Interview with Roy E. Lounsbury, September 7, 1979, *ibid.*

34. *Ibid.*

35. Emergency Operations Center Log, "Assets Available in Addition to those on the Scene 3/31/79," Tab 29, *ibid.*

36. Interview with Roy E. Lounsbury, September 7, 1979, *ibid.;* "Minutes of a Worst Case Contingency Planning Meeting at the State of Pennsylvania Emergency Center in Harrisburg, 31 March 1979," in Emergency Operations Center Log, March 31, 1979, *ibid.*

37. President's Commission, Testimony of L. Joe Deal, August 20, 1979, p. 44, *ibid.*

38. Interview with Richard Beers, October 30, 1979, *ibid.*; interview with Captain C. Creel and Lieutenant R. Bauland, July 27, 1979, *ibid.*

39. Interview with Tom Maguire, July 18, 1979, *ibid.*; interview with William R. Casey, August 14, 1979, *ibid.*; interview with Robert Friess, August 13, 1979, *ibid.*

40. Interview with G. Robert Shipman, July 18, 1979, *ibid.*

41. Emergency Operations Center Log, March 31, 1979, Tab 28, *ibid.*

42. Emergency Operations Center Log, March 31, 1979, Tab 30, *ibid.*

43. Emergency Operations Center Log, Chronology, pp 12-13, *ibid.*

44. *Ibid.*, p. 15.

45. *Ibid.*, p. 16.

46. *New York Times*, April 2, 1979, p. 31.

47. *Ibid.*; see also *Washington Post*, April 18, 1979, p. 16, and "Chronology of Events at INEL related to Three Mile Island," Idaho National Engineering Laboratory File, Three Mile Island Records.

48. *Washington Post*, April 18, 1979, p. 16.

49. Emergency Operations Log, Chronology, pp. 15-21, Three Mile Island Records; Brookhaven Log, pp. 13-14, *ibid.*, Environmental Measurements Laboratory "Radiation Measurements following the Three Mile Island Reactor Accident" (EML-357), May 1979; "Department of Energy Radiological Response to the Three Mile Island Accident," April 14, 1979, pp. A-6, 7; Wayne Adams, tape, no date, Three Mile Island Records; interview with John F. Doyle, October 30, 1979, *ibid.*; interview with Richard Beers, October 30, 1979, *ibid.*

50. Nuclear Regulatory Commission, Closed Commission Meeting, April 2, 1979, pp. 5, 8-9, copy in Nuclear Regulatory Commission File, *ibid.*; Nuclear Regulatory Commission, Press Conference on Three Mile Island, April 2, 1979, pp. 3-5; Nuclear Regulatory Commission, Preliminary Notification-79-67H (April 2, 1979), copies in *ibid.*

51. *Harrisburg Patriot*, April 2, 1979, p. 9 and April 4, 1979, p. 15; *Washington Post*, April 2, 1979, p. A1; *Wall Street Journal*, April 2, 1979, p. A2.

52. Emergency Operations Center Log, Chronology, pp. 21-23, Three Mile Island Records.

53. Walter W. Weyzen to Ruth C. Clusen, April 3, 1979, copy in Emergency Operations Center Log, April 3, 1979, Tab 45, *ibid.*

54. Emergency Operations Center Log, April 2, 1979, Tab 34A, *ibid.*

CHAPTER VI

1. Emergency Operations Center Log, April 3, 1979, Three Mile Island Records; interview with Roy Lounsbury, September 9, 1979, *ibid.*; interview with R. Bauland and C. Creel, July 27, 1979, *ibid.*

2. See Plume Flight Log, April 1-6, 1979, *ibid,*; interview with Jac Watson, July 17, 1979, *ibid.*

3. Idaho National Engineering Laboratory Chronology of Events, March 30 - April 1, 1979, Idaho National Laboratory File, *ibid.*; Vince J. D'Amico to George P. Dix, August 8, 1979, Oak Ridge File, *ibid.*

4. Interview with John M. Doyle, October 3, 1979, *ibid.*; L. Joe Deal, memo to file, August 24, 1979, Environmental Folder, *ibid.*

5. L. Joe Deal, Deposition, August 24, 1979, President's Commission file, *ibid.*; see also, "Suggested Draft Input Memo for Under Secretary," April 11, 1979, Emergency Operations Center Log, Attachment 90, *ibid.*

6. Walter H. Weyzen to Ruth C. Clusen, Memorandum, April 3, 1979, Emergency Operations Center Log, Attachment 45, *ibid.*; interview with G. Robert Shipman, July 18, 1979, *ibid.* interview with Thomas McGuire, July 18, 1979, *ibid.*

7. Situation Report, April 2, 1979, Emergency Operations Center Log, Attachment 40, *ibid.*

8. Nuclear Regulatory Commission, "Preliminary Notice of Events (PNO), PNO-79-671, April 3, 1979," in Emergency Operatons Center Log, Attachment 47, *ibid.*; record of booklet sent in Emergency Operations Center Log, April 2, 1979, *ibid.*

9. "Offsite Population Dose and Risk Assessment from Accident at Three Mile Island Nuclear Station," April 3, 1979, in Emergency Operations Center Log, April 3, 1979, Attachment 47, *ibid.*

10. Situation Report, April 3, 1979, Emergency Operations Center Log, Attachment 48, *ibid.*

11. L. Joe Deal, Deposition, August 24, 1979, pp. 45-46, copy in Environmental File, *ibid.*

12. Friess quoted in the Brookhaven Log, April 3, 1979, Brookhaven File, *ibid.;* interview with Robert Friess, August 13, 1979, *ibid.*

13. Schweller quoted in the Brookhaven Log, April 4, 1979, *ibid.*

14. For attitudes of those at the site see, Brookhaven Log, April 4, 1979, *ibid.;* Friess quoted in *ibid.;* April 4, 1979, and Emergency Operations Center Log, April 4, 1979, *ibid.*

15. Emergency Operations Center Log, April 3, 1979, *ibid.;* Brookhaven Log, April 4, 1979, *ibid.*

16. On McCool's attitude, see Emergency Operations Center Log, March 30, 1979, Attachment 12 and April 1, 1979, Attachment 32, *ibid.*

17. Deal quoted in typescript of telephone conversation with Donald M. Ross, March 31, 1979, Emergency Operations Center Log, Attachment 24, pp. 7-8 *ibid.* and Deal with Whittie J. McCool, March 31, 1979, Attachment 22, *ibid.*

18. Mound attitudes in interview with Clyde W. Taylor and Walt Wallace, September 21, 1979, *ibid.*

19. Memorandum, Argonne Laboratory, August 7, 1979, *passim,* Argonne Laboratory File, *ibid.*

20. Vince J. D'Amico to George P. Dix, August 8, 1979, Oak Ridge File, *ibid.;* W. H. Smith, *et. al.* to R. A. Wynveen, August 7, 1979, Argonne Laboratory file, *ibid.*

21. Interview with Jerry Rhude and Walt Wallace, September 21, 1979, *ibid.;* Environmental Monitoring Laboratory, "Radiation Measurements Following the Three Mile Island Reactor Accident," p. 2, *ibid.*

22. Argonne National Laboratory, Chronology, April 4, 1979, p. 2, Argonne Laboratory File, *ibid.;* interview with Jerry Rhude, September 21, 1979, *ibid.*

23. Interview with Larry Ybarrando and Nicholas Kaufman, August 6, 1979, Idaho National Laboratory File, *ibid.*

24. *Ibid.*

25. *Ibid.*, Idaho National Engineering Laboratory, Chronology, April 3, April 4, 1979, Idaho National Laboratory File, *ibid.* The Idaho Laboratory continued to conduct further semi-scale testing, including a simulation of the Three Mile Island transient. The Laboratory also conducted analyses of samples of Unit 2's core coolant.

26. FBI Memorandum, April 6, 1979, in Emergency Operations Center Log, Attachment 62, *ibid.*; Harold Denton later advised the FBI that no sabotage was involved at Three Mile Island at any time. See FBI Memorandum, April 10, 1979, Emergency Operations Center Log, Attachment 87, *ibid.*

27. Interview with David E. Patterson, September 11, 1979, *ibid.*; Patterson tapes, March 31, 1979, April 4, 1979, *ibid.*; Emergency Operations Center Log, April 4, 1979, April 5, 1979, *ibid.*; Cold shutdown can be defined as the condition when the reactor is subcritical and the average temperature is less than 200° F. Author's interview with John Yoder, March 13, 1979, *ibid.*

28. Emergency Operations Center Log, April 4-6, 1979, *ibid.*; see also, Three Mile Island Incident, Operation Ivory Purpose, Summary, April 7, 1979, Emergency Operations Center Log, Attachment 64, *ibid.*

29. Gage quoted in an unpublished document of the President's Commission, "Report of the Office of Chief Counsel on Emergency Response," p. 73, President's Commission File, *ibid.*

30. *Ibid.*, p. 150; the Las Vegas equipment was fully funded by the Department of Energy, a fact that was not known to the officials at the time.

31. *Ibid.*, pp. 148-150.

32. Author's conversation with Herbert F. Hahn, February 24, 1980; Hahn was present during the exchange; Three Mile Island Incident, Operation Ivory Purpose, April 7, 1979, Emergency Operations Center Log, Attachment 74, Three Mile Island Records.

33. Thomas Gerusky to Joe Deal, April 6, 1979, in Emergency Operations Center Log, Attachment 77, *ibid.*

34. President's Commission, "Report of the Office of Chief

Counsel on Emergency Response," pp. 149-150, President's Commission Folder, *ibid.; see also,* Jack Watson to James Schlesinger, April 13, 1979, *ibid.*

35. Three Mile Island Incident, Operation Ivory Purpose, Summary, April 17, 1979, Emergency Operations Center Log, Attachment 100 and Situation Report, April 14, 1979, Emergency Operations Center Log, Attachment 104, *ibid.*

36. L. Joe Deal to General Joseph K. Bratton, April 17, 1979, Emergency Operations Center Log, Attachment 110, *ibid.;* see also Three Mile Island Incident, Operation Ivory Purpose, Final Update, April 17, 1979, Emergency Operations Center Log, Attachment 114, *ibid.;* John H. Harley to E. W. Bretthauer, May 3, 1979, Environmental Monitoring Laboratory File, *ibid.*

37. L. Joe Deal, Deposition, President's Commission, August 20, 1979, p. 63, President's Commission File, *ibid.;* Herbert F. Hahn, Memo to File, May 1, 1979, Hahn File, *ibid.;* B. H. Greer to G. H. Smith, April 24, 1979, Nuclear Regulatory Commission File, *ibid.;* when the plant has been decontaminated,the second aerial survey will be flown.

38. "Memorandum of Understanding," April 24, 1979, Emergency Operations Center Log, Attachment 110, *ibid.*

39. L. Joe Deal, Deposition, President's Commission, August 20, 1979, pp. 72-73, President's Commission File, *ibid.;* Department of Energy, "Radiological Response to the Three Mile Island Accident," April 14, 1979, p. 10, Environmental File, *ibid.*

40. Interview with Jac Watson, July 19, 1979, *ibid.*

41. *Ibid.;* interview with Thomas Maguire, July 18, 1979, *ibid.;* interview with G. Robert Shipman, July 18, 1979, *ibid.;* interview with John Doyle and Wayne Adams. October 20, 1979, *ibid.*

42. Interview with Wayne Adams and John Doyle, October 30, 1979, *ibid.*

43. Emergency Operations Center Log, April 2, 1979, *ibid.;* Emergency Operations Center Log, April 3, 1979, *ibid.;* Three Mile Island Incident, Operation Ivory Purpose, Summary, April 12, 1979, Emergency Operations Center Log, Attachment 95, *ibid.*

44. Emergency Operations Center Log, April 1, 1979, *ibid.;* L. Joe Deal, Deposition, President's Commission, August 20, 1979, p. 50, President's Commission File, *ibid.* Officials at Brookhaven National Laboratory believed that the decontamination process was Metropolitan Edison's responsibility, even though the Nuclear Regulatory Commission requested the Department of Energy's assistance. Brookhaven Log, April 2, 1979, Brookhaven File, *ibid.*

45. Emergency Operations Center Log, April 1, 1979, *ibid.;* Vallario finished his report on April 6, *ibid.,* April 6, 1979; John McPhee, *The Curve of Binding Energy* (New York, 1976), p. 81; see also, Brookhaven Log, April 3, 1979, Brookhaven File, Three Mile Island Records.

46. E. J. Vallario, "Onsite Meeting," April 5, 1979, Emergency Operations Center Log, Attachment 60, *ibid.*

47. J. W. Crawford, Report of White House Meeting on Three Mile Island Nuclear Plant Situation, March 30, 1979, Energy Technology File, *ibid.*

48. B. W. Washburn to M. P. Norin, May 30, 1979, Energy Technology File, *ibid.;* Washburn felt that the plant operator's attention was centered on other problems and that consequently the loss of feedwater went unnoticed.

49. Emergency Operations Center Log, April 2, 1979, *ibid.*

50. *Ibid.,* April 3, 1979.

51. Three Mile Island Incident, Operation Ivory Purpose, Summary, April 9, 1979, Emergency Operations Center Log, Attachment 79, *ibid.;* transcript of telephone conversation between Ed Patterson and Roy Lounsbury, April 9, 1979, Emergency Operations Center Log, Attachment 76, *ibid.;* Bixby, at the time of this writing, is still directing the Department's efforts at Three Mile Island.

52. Herbert S. Feinroth, Report of Trip to Three Mile Island Nuclear Plant, April 6, 1979, Energy Technology File, *ibid.*

53. Interview with Jack M. Holl, October 17, 1979; the historians' directions helped preserve a large number of important documents that might otherwise have been thrown out.

54. Situation Report, April 11, 1979, Emergency Operations Center Log, Attachment 93, *ibid.*

55. T. A. Werner to A. J. Pressesky, May 21, 1979, Energy Technology File, *ibid.*

56. George H. Ogburn, Jr. to Neal Goldenberg, May 31, 1979, *ibid.*

57. Harry Alter to R. G. Staker, June 6, 1979, *ibid.;* J. J. McClure to Neal Goldenburg, July 2, 1979, *ibid.;* R. H. Kuhnapfel to M. P. Norin, July 19, 1979, *ibid.*

58. Charles O. Tarr to Neal Goldenberg, July 20, 1979, *ibid.;* Benjamin C. Wei to M. P. Norin, August 27, 1979, *ibid.;* George Sherwood to Shelby T. Brewer, August 28, 1979, *ibid.;* Charles O. Tarr to Neal Goldenberg, September 17, 1979, *ibid.* The actual decontamination work was to be carried out by the Bechtel Power Corporation.

59. A. J. Pressesky, Joint Department of Energy/Industry Initiatives, August 17, 1979, *ibid.* Information acquired from the Three Mile Island accident was applied to a similar accident at the Crystal River, Florida, nuclear power plant. The operators were able to prevent core damage when a power failure halted the flow of feedwater into the containment vessel. See *Washington Post,* March 5, 1980.

60. G. W. Cunningham to William J. Dircks, February 5, 1980, Energy Technology File, Three Mile Island Records.

61. "DOE Review of GPU Recommendation of TMI-2 Containment Purging," n.d., attached to Cunningham's letter, *ibid.*

62. Cunningham to William J. Dircks, February 5, 1980, *ibid.* In March 1980, a public education program connected with the planned releases of the krypton was initiated by Herbert Feinroth. Local public officials attended briefings on the plant and the planned controlled releases and the implications for the public's safety. They were also to receive the latest environmental monitoring equipment, be taught how to use it and asked to return to their towns to conduct their own measurements. The idea was that knowledgeable local officials could pass on radiological information effectively and credibly and reduce the fears and concerns of the people in the region. Author's conversations with Herbert F. Hahn, March 10, 1980, and Herbert Feinroth, February 5, 1980.

CHAPTER VII

1. *Department of Energy Radiological Response to the Three Mile Island Accident,* April 14, 1979, copy in Environmental File, Three Mile Island Records; *Department of Energy Environmental Monitoring Response to the Three Mile Island Accident Recommendations,* May 10, 1979, pp. 1-2, 6, *ibid.*

2. Ad Hoc Population Dose Assessment Group, *Population Dose and Health Impact of the Accident at The Three Mile Island Nuclear Station (A preliminary assessment for the period March 28 through April 7, 1979),* May 10, 1979, pp. 2-4, appendices A and B, copy in Environmental File, *ibid.*

3. "Critique of the Three Mile Island Accident Reponse," May 15-16, 1979, p. 4, copy in Environmental File, *ibid.*

4. Vince J. D'Amico to L. Joe Deal, May 18, 1979, *ibid.*

5. NUREG-0600, p. 2.

6. Testimony of John Deutch, Transcript of Proceedings, President's Commission, April 27, 1979, pp. 20, 31-32, 43, Three Mile Island Records.

7. L. Joe Deal, Testimony, *ibid.,* August 20, 1979, pp. 40-49, 68, 70, *ibid.*

8. President's Commission, *Report,* October, 1979, pp. 8, 17.

9. *Ibid.,* pp,. 38-42, 101, 106-107, 126.

10. *Ibid.,* pp. 76-77.

11. President's Commission, "Report of the Chief Counsel on Emergency Preparedness," October 31, 1979, pp. 20-22, copy in President's Commission Folder, Three Mile Island Records.

12. Mitchell Rogovin, et. al., *Three Mile Island: A Report to the Commissioners and to the Public (Vol. 1),* January, 1980, pp. 136-137.

13. *Ibid.,* pp. 131, 137.

160

14. I. C. Bupp and J. C. Derian, *Light Water: How the Nuclear Dream Dissolved,* (New York: Basic Books, 1978), pp. 181-183.

15. Richard G. Hewlett and Frances Duncan, *Atomic Shield: History of the U.S. Atomic Energy Commission, Vol. II 1947/1952,* (Washington, D.C., 1972), pp. 222-260.

ROSTER

Emergency Action Coordinating Team

BRASHEARS, Charles N. . . . . Administration
BRATTON, Joseph K. . . . . Defense Programs
CANNON, James S. . . . . Public Affairs
CULPEPPER, James W. . . . . Defense Programs
KAISER, Ronald L. . . . . Administration
McCOOL, Whittie J. . . . . Environmental Safety
PATTERSON, David E. . . . . Environmental Safety
SHERWOOD, Wayne . . . . Defense Programs
SHULL, Ralph G. . . . . Defense Programs
STUSH, John P. . . . . Administration
VADA, Albert J. . . . . Defense Programs
WEISZ, George . . . . Defense Programs

Emergency Operations Center

BAKER, Kenneth R.
BEAUFAIT, L. J.
BRINKERHOFF, Lorin C.
BROWN, Blake P.
CAVES, Carl A.
COMBS, Gerald L.
DUNCKEL, Thomas L.
EMBER, William N.
GLOVER, N. R.
HILL, James R.
KLEY, George E.
LILLIAN, Daniel
LOOP, Earvin K.
MAYBEE, Walter W.
O'CONNOR, John T.
ROSS, Donald M.
WEINTRAUB, Arnold A.
WELTY, Carl G., Jr.
WOLFF, William F.
YODER, John A.

Office of Energy Research

BECKJORD, Eric S.

Office of Energy Technology

FEINROTH, Herbert
FERGUSON, Robert L.
HAVENSTEIN, Paul L.
KIRKPATRICK, Richard F.
KORNACK, Wallace R.
PRESSESKY, Andrew J.
SMITH, Lloyd A.
TEW, J. Lee
WERNER, Thomas A.

Office of Environmental Compliance and Overview

DEAL, L. Joe
McCRAW, Tommy F.
PATTERSON, David E.
VALLARIO, Edward J.
WEYZEN, W.W.H.

162

Historian's Office
HEWLETT, Richard G.
HOLL, Jack M.

Office of Inertial Fusion
GUE, Marjorie F.
HARMAN, Cathy

Office of Military Applications
LOUNSBURY, Roy E.
PLUMMER, Walter W.

Office of Nuclear Materials Production
ZEOLI, E. Genee

Office of Public Affairs
BRADSHAW, Gail L.
CANNON, James S.
GRIFFIN, James A.
HARRIS, John A., Jr.
LYMAN, James D.
NEWLIN, Robert W.

Brookhaven Office
FRIESS, Robert A.

Pittsburgh Naval Reactors
SAGE, James D.

Schenectady Naval Reactors
SCHOENBERG, Theodore

Bettis Atomic Power Laboratory
BEETLESTONE, Regis C.
BOGARD, Dale
DiNUZZO, Donald M.
ENDLER, Henry P.
FOX, Thomas J.
HANDRA, C. Byron
HOLLIS, Edward
LEHMAN, Paul H.
REED, Thomas E.
SCHULTZ, Boyd G.
SHOTWELL, Merle T.
STEINMETZ, John N.
TIRPAK, John G.
TOBIN, John
WEISFIELD, Michael W.
WROBLEWSKI, Alfred A.
ZEDO, Bernard J.

Brookhaven National Laboratory
BALSAMO, Joseph C.
CARTER, Elbert N.
CASEY, W. Robert
DISTENFELD, Carl H.
GREENHOUSE, Nathaniel A.
HULL, Andrew P.
KLEMISH, Joseph R.
KUEHNER, Alan V.
LESSARD, Edward T.
LEVINE, Gerald S.

LUKAS, Alfred E.
MALINOWSKI, Thomas S.
MILTENBERGER, Robert P.
PHILLIPS, Leigh F.
ZANTOPP, Rudolf

Environmental Measurements Laboratory

BOYLE, Michael C.
GOGOLAK, Carl V.
GULBIN, John F.
MILLER, Kevin M.

Knolls Atomic Power Laboratory

BARNEY, David M.
BROWER, Philip M.
DeVITO, Thomas A., Jr.
EILAND, H. Morris
FOELIX, Charles F.
FOUNTAIN, George R.
GIROUX, Richard R.
HEITKAMP, Albert F., Jr.
MEWHERTER, Jack L.
REKART, Theodore E.
ROBERTSON, Donald B.
STUART, Robert L.
WALKER, Frederick W.
WARD, Winston E.
WINTERS, Ronald W.
YOUNGBLOOD, Jack W.

Mound Laboratory

ANDERSON, James M.
HOPKINS, Lowell C.
PHILLABAUM, Gerald L.
RHUDE, Phillip J.
TAYLOR, Clyde W.
WALLACE, Walter K.

Oak Ridge Office

ALEXANDER, James T.
DAVIS, Bobby Joe
GOODMAN, James W.
LEE, Dave J.
WARREN, Paul E.
WESCOTT, J. Edward

Oak Ridge National Laboratory

BALL, Sy J.
BROOKSBANK, Robert E.
BUTLER, Alfred C.
CAMPBELL, David O.
CARDEN, William D.
CARROLL, Radford M.
CLARK, Roy L.
COLLINS, Edward D.
CONNER, Mitchell L.
COPELAND, Thomas E.
ELDRIDGE, James S.
FRAZIER, Robert W.
FRY, Dwayne N.

HAMLEY, Steve A.

HARRINGTON, Frank E.

"HERMAN"

JOHNSON, William M.

KING, Les J.

KRYTER, Robert C.

PANKRATZ, William L.

POWERS, Bryce A.

SHANNON, William A.

SHEPARD, Robert L.

SIDES, William H.

SMITH, Chris, M.

SMITH, James, E.

SNIDER, James W.

TURNER, Richard E.

YARBRO, Orlan O.

Chicago Office

JASCEWSKY, Edward J.

NEESON, Paul M.

PITCHFORD, Gary L.

Argonne National Laboratory

ELLO, Joseph G.

GAINES, Clifford L.

HUNCKLER, Carl A.

KINSELLA, George E.

MUNDIS, Robert L.

REILLY, David W.

ROBINET, McLouis J.

SEVY, Robert H.

Nevada Operations Office

ADAMS, Wayne . . . . Las Vegas

HAHN, Herbert F. . . . . Andrews Air Force Base

Aerial Measurement System/Nuclear Emergency Search Team
Andrews Air Force Base

AHRENS, Willard E.

BEERS, Richard H.

BLAUVELT, Robert L.

DANIELS, Karen C.

DENNIS, George R.

DINES, Ralph T., Jr.

EICHER, Richard J.

FRANKHAUSER, Walter A.

FREED, Dennis E.

HARRIS, Issac F.

MAGUIRE, Thomas c.

MITCHELL, Janet W.

RIVERO, Arturo Z.

SHIPMAN, Robet T.

WANCHIC, Daniel J.

WATSON, Crestle, Jr.

WATSON, Jac D.

Aerial Measurement System/Nuclear Emergency Search Team
Nevada

ARAMBULA, Leo R.

BEHARY, James G.

BLUITT, Clifton M.

BULL, George H.

BURSON, Zolin G.

CONNERTON, Richard T.

DAHLSTROM, Thomas S.

DAVIS, Stephen G.

DAVISON, Gerald T.

DOYLE, John H.

ENGLISH, Harold R.

GINSBERG, I. William

GLADDEN, Charles A.

JOBST, Joel E.

KNIGHTEN, M. Kirk

LAMONDS, Harold A.

LUPTON, William R., Jr.

MAINA, Lola M.

MEIBAUM, Robert A.

MENKEL, Gary E.

MILLER, Suzi

NAIL, Donald W.

PARELMAN, Martin H.

SASSO, L. Gene

SELPH, Bette J.

SHULTZ, Edward D.

SMITH, Douglas G.

STUART, Travis P.

TICHENOR, Douglas D.

TIPTON, W. John

VIGIL, Frank L.

WINDOM, Rex W.

WINGROVE, Marvin V.

WRIGHT, Jesse J.

EG&G, Inc.—Idaho
BIXBY, Willis W.

KAUFMAN, Nicholas C.

YBARRONDO, Lawrence J.

Lawrence Livermore
Laboratory
DICKERSON, Marvin H.

GUDIKSEN, Paul H.

Savannah River Laboratory
PENDERGAST, Malcolm M.

PENDERGAST, Sherril A.

SCHUBERT, James F.

SYNOPTIC CHRONOLOGY
for
THREE MILE ISLAND

SOURCES OF CITATIONS IN SYNOPTIC CHRONOLOGY

1. Aerial Measuring Systems/Nuclear Emergency Search Team Plume Flight Log

2. Bores, Robert—Notes, Nuclear Regulatory Commission File

3. Brookhaven National Laboratory Communications Log

4. Department of Energy Emergency Operations Center Chronology

5. Department of Energy Radiological Response to Three Mile Island Accident—Log

6. Department of Health, Education and Welfare Chronology

7. Environmental Protection Agency Chronology

8. Hahn, Herbert—Material, Three Mile Island Records

9. Idaho National Engineering Laboratory Chronology

10. Newspaper File
    a. Associated Press Release
    b. *Harrisburg Patriot*
    c. *Houston Post*
    d. *New York Times*
    e. *Wall Street Journal*
    f. *Washington Post*
    g. *Washington Star*

11. Nuclear Regulatory Commission Log

12. NUREG-0600—*"Investigation Into the March 28, 1979 Three Mile Island Accident by the Office of Inspection and Enforcement, U.S. Nuclear Regulatory Commission"*

13. Operation Ivory Purpose File

14. Patterson, David—File, Three Mile Island Records

15. Preliminary Notification of Event or Unusual Occurence (NRC)

16. *"Report of the President's Commission on the Accident at Three Mile Island"* (October 1979)

17. Trip Report—Historian's Office File

SYNOPTIC CHRONOLOGY
for
THREE MILE ISLAND

SYNOPTIC CHRONOLOGY/DEPARTMENT OF ENERGY RESPONSE AT THREE MILE ISLAND

Date/Time	Three Mile Island Event	DOE Response	Other Responses
March 28, 1979 0400	The turbine generator at the Met Ed Three Mile Island nuclear power plant shut down due to a valve failure in the reactor's colling system. (12)		
0654	Because of numerous alarms and area radiation monitors, Metropolitan Edison declared a site emergency. (12)		
0702	Metropolitan Edison notified the Pennsylvania Emergency Management Agency of the emergency at Three Mile Island and asked that the Pennsylvania Bureau of Radiological Health be notified. (10-d)		
0704			Dornsife contacted Margaret Reilly, Chief, Division of Nuclear Reactor Review and Environmental Surveillance, PaBRP, and suggested that she and other staff members report to the office immediately. (16)
0709	Metropolitan Edison officials notified Dauphin County Civil Defense. (12)		
0709-0716	Metropolitan Edison calls the Nuclear Regulatory Commission, Region I at King of Prussia, Pennsylvania. (11)		
0710		From TMI Richard Bensel of Metropolitan Edison called Brookhaven National Laboratory (BNL) alerting them to the problem and the possible need for a Radiological Assistance Team (RAT). Bensel did not request immediate assistance. The BNL RAT went on stand-by status. (3)	

0712		At about this time, Dornsife called the licensee for a status report. The licensee contacted him a few minutes later, but hung up quickly. Dornsife was confused about the status of the plant and possible evacuation of surrounding counties. (16)
0718	The site contacted the Pennsylvania State Police. (12)	
0724	Metropolitan Edison publicly declared a "general emergency" due to higher radiation levels in the reactor containment dome. (12)	
0730	Metropolitan Edison called the Dauphin County Civil Defense to inform them of the general emergency. (12)	
	TMI called PEMA to report the general emergency situation. (12)	
0740	The site phoned NRC Region I of the general emergency. (12)	
March 28, 1979 0745		The secretary at NRC Region I office opened the switchboard and received a message that the licensee had attempted to contact the office to report an emergency. The secretary passed the message to Region I branch chiefs. (11)
		Governor Thornburgh of Pennsylvania informed of accident at TMI. (16)
0750		Thornburgh first thought of ordering an evacuation. (16)
0845	David Schweller, Director of the Brookhaven Area Office (BHO), notified the Department of Energy's Emergency Operations Center (DOE/EOC) in Germantown, Maryland of the problems at TMI. (3)	A five man On Site Inspection Team (OIT) departed NRC Region I office. NRC's team consisted of 1 investigator, 3 health physicians, and 1 reactor inspector. (11)
0850		BNL places two Radiological Assistant Plan (RAP) teams on stand-by alert. In the event of an active response, the United States Coast Guard (USCG) agreed to furnish helicopter transportation to TMI area. (3)

March 28, 1979
0855

DOE/EOC verified the "general emergency" situation with the NRC's operations center in Bethesda, Maryland. (4)

0900

Robert Bores of NRC, Region I, contacted Schweller at the BHO regarding the RAP. At the same time, Charles Meinhold, Director of Health and Safety, BNL, contacted the Pennsylvania Bureau of Radiological Protection with an offer of assistance. (2)

0905

NRC notified the Environmental Protection Agency (EPA) of the TMI incident. Shortly thereafter, the EPA activated its radiation alert office in Washington and placed several of its regional offices on a daily collection schedule. (7)

0915

DOE Emergency Action Coordinating Team (EACT) and staff were notified of accident. (14)

0935

NRC Region I inspector reported that the licensee saw no significant leakage and advocated no evacuation from the site. (12)

March 28, 1979
1000

J. Beaufait of DOE's EOC requested that the Aerial Measuring Systems/Nuclear Emergency Search Team (AMS/NEST) at Andrews Air Force Base (AAFB) go on alert for a possible response to TMI. (4)

Robert Friess (BHO) phoned Bores at NRC Region I to ask if NRC wanted an AMS survey of TMI. Because Bores had no information regarding off-site radiation releases, he said no, "not at this time." Bores also rejected an offer of assistance from a BNL/RAP team. (3)

1002

Metropolitan Edison requested a State Police helicopter to monitor the air over the plant. (10-a)

Friess called DOE/EOC to request that AMS go on stand-by alert. (3)

1010	NRC On Site Inspection Team arrived at TMI. Received briefing on the plant's status. (12)	Bores called BHO to request that AMS go on stand-by. (3)	
1020		Bettis Atomic Power Laboratory (Bettis) placed on stand-by alert. (5)	
1030		The NRC Operations Center in Bethesda requested that AMS go on alert status. (4)	
March 28, 1979 1030			NRC informed the Bureau of Radiological Health (BRH) of the Food and Drug Administration (FDA), Department of Health, Education and Welfare (HEW), of the accident. (6)
1032		Oak Ridge Operations (ORO) went on stand-by alert. (4)	
1035		Friss (BHO) informed NRC that AMS could be at Capital City Airport, New Cumberland, Penn., within two to three hours. (3)	
1045	Utility detected offsite radiation at the level of 3 millirems per hour. (16)	EACT members called for a meeting. (4)	
1055			William Scranton held a press conference. He reported no increase in normal radiation levels outside the plant. (16)
1100		EACT convened in Germantown. (4)	
		Bores (NRC) requested that AMS aircraft from Andrews Air Force Base proceded to Capital City Airport and wait for further instructions. (4)	
		DOE/EOC requested that AMS and RAP activities be coordinated through the EOC. (4)	
March 28, 1979 1100-1400			Two carloads of NRC officials left headquarters in Bethesda to team up with Region I van at TMI. Upon arrival, they discovered that TMI phones were jammed. (10-f)

1115		NRC Operations Center told BHO that a RAP team was not needed at this time. (3)
1118		Margaret Reilly, PaBRP, requested a DOE/RAP team from BHO. (3)
1130		EACT requested that AMS team proceed from Andrews Air Force Base to Capital City Airport. (13)
1130-1800	NRC monitoring radiation problems. (11)	
1225		AMS team left Andrews Air Force Base for Capital City Airport on an Air Force helicopter. Herb Hahn, Bob Shipman, and Ike Harris constituted the advance party. (4)
1245		Pennsylvania State Police blocked Route 441 to traffic, upon request from PaBRP. (12)
1330		A second helicopter, this one loaded with aerial measuring equipment, left Andrews for Capital City Airport. (4)
		Air Force helicopter with advance AMS/NEST party landed at Capital City Airport. Hahn established communications with DOE/EOC. (4)
March 28, 1979 1350	Pressure spike of 28 psig felt in Unit 2 control room, resulting from a hydrogen burn in the reactor building. (12)	
1400		DOE/EOC confirmed that BNL/RAP team would be working for Pennsylvania and the AMS/NEST team would be working for NRC. (2)
1400-1430		An NRC inspector made surveys in Harrisburg and found no radiation levels above those of background. (12)
1410		Atmospheric Release Advisory Capability (ARAC) Center gave weather conditions. (8)

1430	AMS/NEST helicopter reported first radiation reading in the plume above TMI. (14)	
	NRC informed Idaho National Engineering Laboratory (INEL) of incident at TMI. (9)	
1439	Coast Guard helicopter with BNL/RAP team and air and sail sampling instruments arrived at Capital City Airport. (3)	
1445	The second AMS/NEST helicopter with instruments aboard landed at Capital City Airport. It immediately refueled and launched the first aerial survey of the radiation levels of airborne releases from the TMI power plant. (13)	
March 28, 1979 1500	Herb Hahn established the DOE Command Post in the manager's office at the Capital City Airport. (11)	
1525	BNL/RAP team began to monitor milk, vegetation, and air samples northwest of the plant. (4)	
1545-1653	DOE/EOC staff briefed Congressional and other DOE staff on the situation at TMI. (4)	
1559		NRC senior manager communicated to licensee over telephone NRC's concern that pressure level indicator did not preclude a bubble in the core. Core may be uncovered. (12)
1600-1700	AMS plume flight which showed a reading of 0.2 milliroentgens per hour (mr/hr) seven miles from the plant. AMS personnel then relayed this information to NRC, Region I. (1)	
1610		An NRC Region I inspector reported that licensee believed that there was no bubble in the core. (12)
1630	An NRC inspector in the Unit 2 Control Room reported that bubbles in the cooling system had been removed. (12)	

March 28, 1979 1656		An NRC inspector reported to Region I that licensee concluded that core was uncovered. (12)
1710	D. E. Patterson selected as senior DOE representative at TMI to coordinate DOE activities with the State of Pennsylvania and NRC. (4)	
1815-1830	DOE/EOC orders Bettis to upgrade its RAP team status from stand-by alert to full alert. Argonne National Laboratory (ANL) placed on stand-by alert. (4)	
1845	EG&G monitoring teams from Las Vegas left for Capital City Airport. (8)	
1900	Colonel Roy Lounsbury (DOE/EOC) contacted the office of the Lt. Governor of Pennsylvania to explain the DOE's role in responding to the incident at TMI. (4)	
1910-2000	Bores (NRC Region I) requested an AMS plume flight which was then launched. Radiation levels proved to be lower than those found during the 1600-1700 flight. (2)	
2030	Bob Friess (BHO) briefed Lt. Governor William Scranton III and his staff on DOE activities. (8)	
2130	Friess (BHO), Lt. Governor Scranton, NRC, and Thomas Gerusky of PaBRP hold a press conference in Harrisburg. (8)	
March 28, 1979 2200		Dr. Charles Gallina, and NRC vestigator, said that the worst was probably over. "The reactor is stable," said Gallina. (10-d)
2230	Hahn reported to the DOE/EOC that radioactivity was escaping from the reactor building into the auxiliary building. (4)	
2300	Friess and NRC turned AMS and RAP team data to Governor Richard Thornburgh of Pennsylvania. (8)	
2316	Ed Patterson arrived at Capital City Airport. (4)	

174

March 29, 1979
0200

A second BNL/RAP Team arrived. (8)

0210

Hahn informed DOE/EOC that a second NEST team had departed from Andrews Air Force Base bound for Capital City Airport. (4)

0400

Group from AMS/Andrews Air Force Base arrived with a data van, search van, and portable analyzers. (8)

0435

The makeup tank was vented to the vent header to the waste gas decay tanks. Operations personnel knew that the vent header leaked and that venting to the vent header resulted in releases to the auxiliary building exhaust system. (12)

0700

A plume flight gave a reading of 0.1 mr/hr at a distance of eight miles from the plant. (1)

0846

Jim Sage, Pittsburgh Naval Reactors (PNR) placed Bettis RAP team on full alert with personnel and equipment ready. (3)

0937

A plume flight took place in response to a probable release of radiation from the plant. The flight spotted xenon 133 as well as some radiation leakage from the containment building. (4)

1000

Patterson, DOE Command Post, requested that Bettis send its RAP team. (13)

1000-1100

AMS/NEST flew a plume flight to measure the size of the plume. No radiation levels were measured during this flight. (1)

March 29, 1979
1030

Las Vegas AMS team arrived at Capital City Airport. (8)

1032			NRC told the licensee that no iodine appeared in the Unit 2 control room air samples collected by the licensee. (11)
1215	Floor in the auxiliary building now covered with plastic sheeting to reduce release rate of radioactive gases from water on the floor. (12)		
1223			Scranton went to the Observation Center for a briefing with Metropolitan Edison. Subsequently, he was allowed in the auxiliary building of TMI for an inspection. Scranton said that "there is no cause for alarm." (11)
DURING THE AFTERNOON			Thornburgh said that the situation did not merit evacuation of pregnant women. (10-d)
1400		L. Joe Deal left Germantown for Command Post at Capital City Airport. (4)	
1410	A Metropolitan Edison helicopter measured 3,000 mr/hr beta-gamma and 400 mr/hr gamma at 15 feet above the stack. (12)		
March 29, 1979 1410-1815	Metropolitan Edison dumped 25,000 gallons of radioactive waste water into the river. (12)		
1430			Bores called Gerusky and found out that Xe133 and Xe135 was in the water at TMI. Gerusky was in favor of dumping the 400,000 gallons of water. (2)
1510		One BNL RAP team departed from Harrisburg International Airport. A second BNL RAP team remained in order to brief their replacements from Bettis. (4)	
1600		The Bettis RAP Team arrived at Capital City Airport. They immediately left with Ed Patterson for Harrisburg to be briefed by BNL's remaining RAP team. (4)	
1600-1700		A plume flight occurred which showed a reading of 0.5 mr/hr at 500 feet altitude, one half mile from the plant. (1)	

March 29, 1979 LATE AFTERNOON		Thornburgh, Gerusky, and Dr. Charles Gallina of NRC held a press conference. Everyone agreed that no danger existed for area residents. Gerusky said that radiation levels near the site had decreased from 20 mr/hr at 0630 to 1 mr/hr at 2 p.m. Charles Gallina said that conditions in the area were "steadily improving." (10-b)
1755		NRC's Executive Management Team directed the licensee to stop dumping all water. Region I notified NRC HQ that stopping the dumping would cause water to back up in the turbine building. (11)
1800		PaBRP said that Metropolitan Edison could dump the water. NRC could make the final decision without further consulting PaBRP. (11)
1830		NRC learns results of an analysis of a primary coolant sample: damage to the core was far more substantial than had been expected. (16)
March 29, 1979 DURING THE DAY		Hendrie and NRC staff testified before House Subcommittee on Energy and Environment, chaired by Representative Morris Udall. Hendrie said that the problem at TMI was caused by a series of mechanical failures. (10-d)
1845	Sage and Bettis RAP team arrived at the offices of PaBRP in Harrisburg to be briefed by BNL RAP team. (3)	
2000	Joe Deal arrived at the Command Post at Capital City Airport. (8)	
2045	Sage called Friess at home in New York in order to receive a briefing and all relevant data taken by BNL teams. (3)	
2200	A second RAP team from BNL left for TMI by truck. (3)	
2230-2330	Plume flight readings showed 0.5 mr/hr at 500 feet altitude, on half mile from the plant. (1)	

March 30, 1979
0130

A Metropolitan Edison Nuclear engineer said that he notified the State of Pennsylvania and the NRC representative at TMI of a planned venting of a tank around 0200-0300. He also said that he followed this procedure each time he knew of a vent. (11)

0150-0350   Metropolitan Edison vented the make-up tank for short intervals of time. (12)

0700

0744-0800   Licensee notified NRC representatives at plant and Pennsylvania officials that make-up tank had been vented and that increased releases were expected. Later, the licensee called back and said that releases would be greater than originally anticipated. (11)

0756   Metropolitan Edison helicopter measured 1,000 mr reading 130 feet above the Unit 2 auxiliary building. (12)

0800

Joe Deal informed the DOE/EOC that although the reactor was venting some radioactivity, the releases were under control. (4)

0801   Metropolitan Edison helicopter measured 1,200 mr/hr reading at same position. (12)

March 30, 1979
0830-0900   Metropolitan Edison notified PEMA of new releases from gas decay tanks. (11)

E. C. McCabe of NRC at Trailer City told reporters of high radiation reading in the air. (10-f)

0834   Licensee told Pennsylvania Civil Defense to relay the message of new releases to PaBRP. Some confusion existed as to whether the licensee recommended evacuation of the site. (11)

0900

Just after this time, Carter called Hendrie. Hendrie wanted better communications. The White House Signal Corps installed telephone lines to hook up the White House, State of Pa. offices, and the control room in four hours. (10-f)

0925		NRC Region I called the state about new releases. Despite rumors, the site did not plan to call for evacuation. (11)
0928	Because of an anticipated planned release of radioactivity, BHO and PNR discussed the possibility of relieving Bettis' RAP teams with BNL personnel and vice versa. Equipment would be shared and all Bettis and BNL monitoring would be coordinated through Joe Deal. (3)	
March 30, 1979 1000		Paul Critchlow, Press Aide to Governor Thornburgh, issued a statement from the Governor advising those within a 10 mile radius of the plant to stay indoors and to keep the windows shut. (10-b)
		Grier called Gerusky and said that his office (Region I) did not recommend evacuation. (2)
1015		Grier called Gerusky after having conferred with NRC HQ (Norman Mosely) and confirmed that, at that point, the NRC did not recommend evacuation. Gerusky said that the state was recommending that people stay indoors. (2)
1025		NRC Region I lost contact with the Unit I control room. (11)
1030		NRC Region I continued to recommend no evacuation. (11)
March 30, 1979 1030-1130	A plume flight gave a reading of 20-30 mr/hr at 300 feet altitude, 1/4 mile southwest of the plant. (1)	
1045	In response to an NRC request, BNL sent a meter capable of reading 1000 R/hr. (3)	
1100	An unplanned release of a significant quantity of radioactive material into the atmosphere was reported. Local residents were told to take shelter. (3)	

1115		Carter called Thornburgh to discuss the effect of the new releases. The President said that he would send Denton, a special army communications system, and antiradiation medicine to TMI. He remarked that it is "best to err on the side of safety and caution." (10-d)
		Air raid sirens in Harrisburg went off, causing some panic in the city. (10-f)
1130	In response to an NRC request, DOE/EOC called in RAP teams from Oak Ridge National Laboratory (ORNL) and Argonne National Laboratory (ANL) to aid in the monitoring efforts. (4)	At about this time, Hendrie and NRC personnel recommended a partial evacuation to Thornburgh. (10-d)
March 30, 1979 1149	DOE/EOC placed two BNL RAP teams on stand-by alert to relieve other RAP teams as needed. (3)	
1200-1300		Governor Thornburgh held a press briefing. Thomas Gerusky was also present. The governor ordered pregnant women and small children to evacuate a five mile radius of the plant. Thornburgh also placed those within a ten mile radius of the plant (about 140,000 persons) on alert and closed area schools. "We cannot predict what the situation will be in 24 hours," remarked Thornburgh. (10-b)
1230	Metropolitan Edison requested ORNL personnel with specific capability to monitor iodine which was believed to have been released during the morning. (4)	
1245	The AMS/NEST H-500 helicopter grounded due to engine trouble. Hahn requested additional equipment and personnel from 1st Helicopter Squadron at Andrews Air Force Base. (4)	
1330		Saul Levine of NRC called the Idaho Engineering Laboratory and EG&G for assistance in dealing with the hydrogen bubble. (9)

March 30, 1979
1330

The White House held a
meeting with DOE, DOD,
FDAA, NRC, and other
federal agencies regarding
the problems at TMI. (4)

1400

DOE/EOC requested that
Lawrence Livernore
Laboratory (LLL) send
Paul Gudiksen to relieve
Marv Dickerson of the
ARAC operation. (4)

Harold Denton and 12
other NRC personnel ar-
rived at the site. Denton
said that he planned to
work closely with the
Governor. Shortly there-
after, the National Security
Council called. The President wanted to talk to Denton in order to get a status
update. Denton left a
meeting with John Herbein,
Vice-President of Metro-
politan Edison, in order to
speak with the President.
(11)

1450

NRC informed DOE/EOC
of intermittent radioactive
releases from the primary
coolant system. Evidence
indicated that there had
been severe fuel damage
and the presence of a large
hydrogen gas bubble in the
reactor vessel. (4)

March 30, 1979
1515

Region V RAP coordinator
and three others from ANL
left Chicago by plane for
TMI. (13)

1600-1800

A plume flight indicated
reading of 9 mr/hr, one
mile from the plant. (1)

1650

ORO informed DOE/EOC
that a six man RAP team
would be enroute to TMI.
Two additional men would
drive up on the 31st bringing 25-50 respirators. (4)

1830

DOE's Nevada Operations
Office (NVOO) told
DOE/EOC that it would
send two communication
pods on April 1 from Las
Vegas and two more on
the same day from Los
Angeles. (4)

2000

ANL RAP team arrived at
Capital City Airport.
Another RAP team de-
parted from BNL for
Capital City Airport. (3)

2005	Two Air Force helicopters left Andrews Air Force Base for TMI. (4)	
2100	ORNL RAP team arrived at Capital City Airport. (13)	
2120-2225	Plume flight measurements indicated a reading of 0.5 mr/hr at 500 feet altitude, one mile from the plant. (4)	
2138	Joe Deal met with NRC and officials of other responding federal agencies. This meeting was the first of what would later be called the "5 o'clock briefings." (4)	
March 30, 1979 2215	BNL RAP team arrived at Capital City Airport. (3)	
DURING THE DAY		Jack Waston, Presidential Assistant for Intergovernmental Affairs, ordered the FDA to ship 240,000 vials of potassium iodide, which protects against radioactive iodine, to Harrisburg. (10-f)
March 31, 1979 0015-0105	Midnight plume flight flown by Air Force crew from Andrews Air Force Base found a reading of 1 mr/hr at 500 feet altitude, one mile from the plant. (1)	
0300		The FDA authorized Mallinckrodt Chemical to begin production of potassium iodide. (10-d)
0300-0400	Scheduled plume flight indicated a reading of 1.5 mr/hr at 500 feet altitude, one mile from the plant. (1)	
0320	NRC requested that DOE locate and send as many lead bricks as were available to expedite a plan to get rid of the hydrogen bubble. (4)	
0415	NRC requested lead bricks from BNL. (3)	
0430	BHO informed NRC that they could deliver some bricks to TMI within twelve hours. (3)	

0505	Bettis informed the DOE/EOC that they had about 1,000 lead bricks two miles from the Allegheny Airport in Pittsburgh in case quick delivery was imperative. (4)
0508	NRC informed the DOE/EOC that approximately 2,000 bricks would be required. (4)
0525	NRC asked Bettis to bring their lead bricks to the Allegheny Airport where an Air Force C-130 would pick them up. Bettis informed the NRC that a second count revealed that 600 bricks were available. (4)

March 31, 1979

0600	ANL monitoring van arrived at Capital City Airport. (13)
0600-0715	Plume flight monitoring equipment measured 2.5 mr/hr at 500 feet altitude, one mile from the plant. (1)
0635	The NRC called Colonel Roy Lounsbury at the DOE/EOC in inquire about DOE planning for an escalated emergency. Lounsbury outlined means of additional assistance, but made no specific arrangements. (4)
0655	The NRC checked with DOE/EOC regarding contingency plans for a "worst-case" event and consequent evacuation. (4)
0715	BNL prepared two flatbed trucks of lead bricks to sent to MacArthur Field in Islip, New York for reshipment to TMI by military aircraft. (3)
0745	The plant manager at TMI made arrangements with the DOE Command Post at Capital City Airport to ship a gas sample from the reactor vessel to Bettis for analysis. (4)
0750	Joe Deal requested the assistance of Colonel Lounsbury at TMI. (4)
0900-1000	Scheduled plume flight indicated a measurement of 3 mr/hr at 500 feet altitude, one mile from the plant. (1)

1020		EG&G/Nevada agreed to send four communications pods to Philadelphia. From there trucks could haul the pods to Capital City Airport. (4)
1030		DOE/EOC requested four U.S. Army trucks with drivers to deliver the communications pods to Capital City Airport. (4)
March 31, 1979 1128-1221		DOE/EOC, BNL personnel, and Bettis scientists discussed the status of the primary core coolant sample. They agreed that the high radiation level of the sample, 100R/hr per 100 ml, was too hot to handle in most laboratories. To ensure adequate radiation protect, the sample would be shipped in a 55-gallon drum. (3)
1200		NRC accepted the responsibility for packaging an shipping the primary coolant sample. (4)
		HEW Secretary Califano, in a memorandum to the White House, summarized HEW activities and recommended that the White House encourage the NRC to consult HEW and EPA public health experts. (6)
1215-1330		A plume flight which registered a reading of 1.5 mr/hr at 500 feet altitude, one mile from the plant. (1)
1225	Reactor status report stated that an off-site radiation release, probably from a leaking relief valve, was occurring. (4)	
1230		Primary core coolant sample was to be analyzed by Bettis and Knolls Atomic Power Laboratory personnel at Bettis' laboratory in Pittsburgh. (4)
1315		Friess reported to BHO that the size of the hydrogen bubble in the reactor core had shrunk by one-third. (3)
March 31, 1979 1335		DOE/EOC briefed Secretary James Schlesinger. (4)
1421		BNL RAP team at site requested specialized portable monitoring equipment from EML. (3)

1427		DOE/EOC asked BHO to provide a list of medical personnel to respond to the scene if needed. (3)
1440	Metropolitan Edison calculated the size of the hydrogen bubble at 880 cubic feet. (15)	
1520-1540		A plume flight recorded 3 mr/hr at 500 feet altitude, one mile from the plant. (1)
1617		BHO reported that between eight and ten doctors were available to aid radiation victims in case of an emergency. (3)
1620	Metropolitan Edison calculated the size of the hydrogen bubble at 621 cubic feet. Two samples of containment atmosphere showed hydrogen concentrations at 1-7 and 1-0%. (15)	
1632		Additional personnel from EG&G/Nevada left for Harrisburg. (4)
DURING THE AFTERNOON		Stuart Eizenstadt, Presidential Aide, called Carter in Wisconsin. Eizenstadt suggested that Carter go to TMI. The President checked with Denton to make sure that the visit would be okay with NRC. (10-f)
March 31, 1979 1800		A RAP team from ANL began operation at a monitoring site in Goldsboro. (13)
1843-1945		Plume flight mission indicated a reading of 3 mr/hr at 500 feet altitude, one mile from the plant. (1)
1925		Colonel Lounsbury requested space to house DOE personnel at the Army's Carlisle Barracks. (4)
1927		The Army Operations Center at the Pentagon told the Carlisle Barracks to support the DOE request. (4)
2100-2145		AMS/NEST plume flight indicated a radiation reading of 0.8 mr/hr one mile from the plant. (1)

2249	The Bettis Laboratory reported levels of Xenon and Iodine in the sample of gas taken from the reactor building. This information went to the DOE/EOC and then to the command post at Capital City Airport. (4)	
EVENING		President Carter, in a speech in Milwaukee, announced that he would go to TMI. (10-f)
April 1, 1979 0030-0100	AMS plume flight indicated a reading of 1 mr/hr at 500 feet altitude, one mile from the plant. (1)	
0155-0245	A special plume flight was to be flown during a sampling of holding tank water. Metropolitan Edison did not take this sample. Radiation levels reached 1 mr/hr at 500 feet altitude, one mile from the plant. (1)	
0340	A Bettis analysis of the primary core coolant indicated no significant fuel melting. (4)	
0600-0700	After holding tank sample was taken, a plume flight indicated a reading of 2-3 mr/hr at 500 feet, one mile from the plant. (1)	
0610	An announced radioactive release from the TMI waste gas decay tank occurred. (4)	
0635	25 additional EG&G staff arrived. (4)	
0805	BNL personnel on site requested an analytical trailer from EML. (3)	
0900-0940	Plume flight indicated a radiation reading of 3 mr/hr at 500 feet altitude one mile from plant. (1)	
0945	DOE/EOC learned that NASA was sending an H2 gas expert at Metropolitan Edison's request. DOE agreed to provide transportation and communication support. (4)	
April 1, 1979 DURING THE MORNING		President Carter and his wife Rosalynn landed at the Air National Guard Facility in Middletown. Governor Thornburgh and Harold Denton, who briefed Carter later, met the President. (10-f)

1200		NRC briefed officials from HEW, EPA, and the White House. (6)
1220	Metropolitan Edison and Electric Power Research Institute (EPRI) requested chemical engineers to aid in cleaning up gaseous and liquid wastes. ORNL agreed to send three men. (4)	
1300	As of 1300 hours on April 1, DOE Support personnel totaled 80. (4)	
1300-1346	Plume flight indicated a reading of 1 mr/hr at 500 feet altitude, one mile from the plant. (1)	
1425-1427	President Carter was on site in the Unit 2 control room. (11)	
1400-1500	For 36 minutes during this time, Carter toured the TMI plant. (10-f)	
1500		President Carter held a press conference in the gymnasium of the Middletown Borough Hall. (10-f)
1800-1900	Plume flight indicated a reading of 0.5 mr/hr at 500 feet altitude, one mile from the plant. (1)	
April 1, 1979		Lee Gossick, NRC Executive Director of Operations, called DOE's Emergency Operations Center (EOC) to request a cleanup of the auxiliary building. General Public Utilities (GPU) also requested help. (11)
2340	Deal arranged a meeting with plant and NRC staff for noon, April 2, on decontamination of the Auxiliary building. (4)	
DURING THE EVENING		Governor Thornburgh issued a statement which said that only those schools within a 5 mile radius of TMI should remain closed. He also directed all state workers to report to work as usual. (10-b)
DURING THE DAY	All DOE/RAT teams became linked by radio. Mound personnel assisted FDA with setting up transluminescent dosimeters (TLDs). (13)	HEW transported the 200,000 remaining information leaflets on potassium iodide to Harrisburg. (6)

April 2, 1979 0300-0400	Plume flight indicated a reading of 1.5 mr/hr at 500 feet altitude, one mile from the plant. Sampling of holding tanks did not change the reading. (1)	
0600-0615	Plume flight during bad weather and poor visibility indicated radiation of 1.0-1.5 mr/hr. (1)	
0812	DOE/CP asked EOC to make arrangements to fly 55 gallon drum of primary core coolant to Bettis in Pittsburgh for analysis. (4)	
1050	BHO requested copies of "Emergency Handling of Radiation Accident Cases" in conjunction with possible evacuation plans. Booklet made available. (4)	
DURING THE MORNING		Denton in a press conference at the Middletown Gym, said that the size of the bubble had decreased dramatically and that the temperature in the reactor had gone down. (10-f)
1200	NRC bulletin admitted that "the net oxygen generation rate inside the noncondensible bubble in the reactor is much less than originally conservatively estimated." (15)	
1310		Herb Feinroth (ET) suggested that Savannah River Laboratory (SRL) personnel be used in decontaminating buildings and equipment at TMI. (4)
April 2, 1979 1320		Friess reported NRC's request that Savannah River Laboratory (SRL) personnel be used in decontaminating buildings and equipment at TMI. (4)
DURING THE DAY		Thornburgh decided to continue earlier directive about pregnant women and young children remaining outside a five mile radius of TMI. (16)
2321		C-141 was on standby to transport the primary core coolant sample at NRC's request. (4)

April 3, 1979

0035		NRC requested DOE assistance in obtaining bladder tanks to store excess contaminated water. (4)
0045		NRC requested a medical doctor to discuss iodine radiation problems. Doctor was to be at NRC trailer at TMI at 0800, April 4. Dr. Weyzen was contacted. (4)
0155		DOE/EOC learned that the FDAA (Federal Disaster Assistance Administration) would handle bladder tank problem for NRC. DOE would not be involved. (4)
0700	Hydrogen recombiner operating. Hydrogen concentration was about 1.9%. (15)	
0905-1005		Plume flight indicated a radiation reading of 9.8 mr/hr at 500 feet, one mile from plant. (1)
0915		Iodine levels were below minimum detectable levels. Samples were to be sent to outside laboratories for analysis. (3)
1000		NRC requested a DOE photographer to document NRC activities at the reactor site. (4)
1200-1245		Plume flight indicated a radiation reading of 0.8 mr/hr at 500 feet altitude, one mile from the plant. (1)
1430		BNL discussed withdrawing personnel as emergency phase had passed. (3)

April 3, 1979

1500-1600		A plume flight indicated a reading of 2 mr/hr at 500 feet, 1/2 mile from the plant. (1)
1515		DOE CP requested Gene Start, National Oceanic and Atmospheric Administration (NOAA) in Idaho Falls, Idaho, to assist ARAC on the scene. (4)
1600		DOE/EOC was authorizing Bettis to analyze primary core coolant sample. DOE/EOC believed that DOE environmental monitoring support was excessive for the present situation. (4)

DURING THE DAY			Congressman Udall called for an investigation of TMI. (10-d)
			Officials in Harrisburg received the remaining shipments of potassium iodide, bringing the total number of bottles delivered to the Harrisburg area to 237,013. (6)
			HEW made preparations to send 80 nurse/physician teams to evacuation centers, if needed. (6)
LATE AFTERNOON			Lieut. Governor Scranton visited the Dauphin County Office of Emergency Preparedness. He urged all officials to remain on alert. (10-b)
April 3, 1979 1715			NRC dubbed TMI operation "Ivory Purpose." (4)
1815-1900		Plume flight indicated a reading of 2 mr/hr at 500 feet, one mile from the plant. (1)	
2100-2145		Plume flight indicated a reading of 0.5 mr/hr at 500 feet, one mile from plant. (1)	
April 4, 1979 0000-0100		Plume flight indicated a reading of 1.1 mr/hr at 500 feet altitude, one mile from the plant. (1)	
0300-0330		Plume flight indicated a reading of 0.3 mr/hr at 500 feet altitude, one mile from the plant. (1)	
0600-0640		Plume flight indicating a level of 1 to 1.2 mr/hr at 500 feet altitude, one mile from the plant, was flown. (1)	
0700	Hydrogen concentration in containment at about 2.1%; hydrogen recombiner in operation. (15)	BHO representative at TMI reported some organizational problems. He suggested that the RAP team response should be terminated as there was no danger to the health and safety of the people near the plant. (3)	
0845		DOE/EOC began planning for reduced levels of monitoring support. The NRC requested a reduced AMS/NEST flight schedule. (4)	

0900-0945	Plume flight indicated a reading of 1 mr/hr at 500 feet altitude, one mile from the plant. (1)
0920	BHO discussed a reduction of personnel at TMI with DOE/EOC. The EOC said that BHO must get prior approval from the State of Pennsylvania for a reduction. (3)
0945	Assistant Secretary Clusen cancelled a scheduled trip to TMI. (4)
1030-1115	Plume flight reflown due to the insensitivity of the instrumentation during the 0900 flight Plume size was monitored, but no additional radiation measurements were taken. (1)
April 4, 1979 1130	ORO agreed to send two chemists to TMI to aid at the site. (4)
1200-1230	Plume flight indicated a reading of 0.7 mr/hr at 500 feet altitude, one mile from the plant. (1)
1300	The NRC requested help in locating contractor laboratories other than Bettis to analyze primary core coolant samples. At the same time, DOE/EOC informed the NRC of the planned phase-down of DOE personnel at TMI. (4)
1345	The NRC requested ORNL to aid Babcock & Wilcox in the evaluation of core conditions. (4)
1515	The DOE Comand Post requested that the Army permit its truck containing the communications pods to remain at Capital City Airport for five more days. (4)
1522-1545	Poor visibility at this time shortened this plume flight. the flight was made to test the Microwave Ranging System. (1)
1535	Metropolitan Edison requested a portable hot cell from DOE. (4)
1550	BHO site representative proposed to reduce current DOE and contractor personnel doing environmental monitoring for Pennsylvania from 29 to 12. (3)

1610		BHO informed Joe Deal of the possibility of ending the RAP response when the reactor reached the cold shut down stage. (3)
1615		The Army granted the extension on the use of its trucks. (4)
April 4, 1979 1620		The DOE informed Metropolitan Edison that it had no portable hot cell. (4)
1830		The DOE Command Post scheduled a reduction in the RAP team levels. As of April 6, all CHO and ORO personnel would be released; 3 people from BNL, 2 from EML., 2 from Bettis, and one from SNR would remain. (4)
April 5, 1979 0600-0645		Plume flight indicated a reading of 0.3 mr/hr at 500 feet altitude, one mile from the plant. (1)
0700	Hydrogen in containment at about 2%; hydrogen recombiner in operation. (15)	
	0837	Herb Hahn informed the DOE/EOC that 102 DOE and DOE contractor personnel were at TMI. (4)
0950-1040		Plume flight readings indicated a level of 0.3 mr/hr at 500 feet altitude, one mile from the plant. (1)
1000-1530		At about this time, Robert L. Ferguson, Herbert Feinroth, Erick Beckjord, Andrew Pressesky, Richard Hewlett, and Jack Holl from DOE headquarters toured the TMI site and Command Post. (17)
1205		Joe Deal requested that team individuals not discuss radiation levels or operations with news media representatives during the phase-down operation. (4)
1223-1327		Plume flight readings indicated a level of 0.1 mr/hr at 500 feet altitude, one mile out. (1)
1400		The DOE Command Post reported 80 DOE and DOE contractor personnel remained at TMI. (4)

1430 — The DOE filed a cost summary of RAP team support to the NRC and the State of Pennsylvania; $819,476 for the period between March 28 and April 5, 1979. (4)

1515-1615 — Plume flight readings showed less than 0.1 mr/hr at 500 feet altitude, one mile from the plant. (1)

April 5, 1979
1645 — The DOE Command Post filed a reduced operation plan with the DOE/EOC to reduce the number of DOE personnel to 42 over the next four days. (4)

1649-1831 — A special plume flight was launched in conjunction with a transfer procedure at the plant. Less than 0.2 rm/hr was detected near the plant. (1)

1750 — Thomas Gerusky of the State of Pennsylvania and Bernie Weiss of the NRC agreed to the DOE phase-down plan. (4)

DURING THE DAY — Joseph A. Califano, Secretary of HEW, announced a long term health study of workers at TMI and pregnant women in the area. He predicted that no cancer death would result from TMI. (10-f)

Carter established a Presidential Commission to study TMI. (10-d)

Using measurements from dosimeters in the area around TMI, a team from HEW, EPA, and NRC estimated that the external radiation doses received by the population around TMI would be relatively low. (6)

State of Pennsylvania and Metropolitan Edison refused an offer for more radiation medication. (10-d)

HEW, NRC, PaDER, Pa. Dept. of Health, and Metropolitan Edison met in Middletown to discuss the exposure of Metropolitan Edison personnel to radiation. (6)

EPA began a milk sampling program, which was designed to complement the programs of the FDA, State of Pa., and Metropolitan Edison. (7)

April 5, 1979
2055 — The NRC asked the DOE/EOC if one of the national laboratories might have five constant flow air meters to detect a high level of iodine. The NRC also requested the same instruments from ORO simultaneously. (4)

2120-2150 — Plume flight indicated levels of 0.05 mr/hr at 500 feet altitude, one mile from the plant. (1)

2325-2345		A special plume flight was requested to monitor a transfer at the plant. When the transfer was delayed, the flight was scrubbed. (1)
April 6, 1979 0545-0630	Waste Gas Decay Tanks were vented, and then closed, when radiation levels at the auxillary building exhaust monitor increased ten-fold. (15)	
0600-0710		Plume flight launched during a transfer of radioactive water at the plant. In very strong winds, the instruments, showed a 0.15 mr/hr reading at 500 feet altitude, one mile from the plant. (1)
0745		The DOE Command Post at Capital City Airport requested approval for a staff reduction from the DOE/EOC. (4)
0900		BNL agreed to send a new RAP team to TMI to relieve present personnel at the site. (3)
0915	Venting of Waste Gas Decay Tanks to the containment building resumed. This increased hydrogen concentration to slightly over 2%. (15)	
1030		Joe Deal met with Harold Denton regarding staffing levels. (4)
1120		Bettis agreed to remain on standby in order to analyze primary core coolant samples over the weekend. (4)
1245		Denton and the NRC approved the DOE personnel reduction plan. The DOE Command Post began to implement the reduction. (4)
1425		The NRC requested that the EOE/EOC arrange for SRL expertise on charcoal-iodine removal. (4)
1640		The NRC asked the DOE/EOC to help with AMS/NEST and ARAC technology information in preparing congressional testimony.

194

April 6, 1979
DURING THE
DAY

Plant discharged waste into
the Susquehanna River.
(10-g)

PEMA estimated that 90%
of those who fled the
nuclear accident had
returned home. (10-g)

Nuclear Waste from TMI
trucked through Virginia on
the way to a South Carol-
ina storage site. (10-f)

The EPA discontinued
drinking water sampling,
except for major sources of
public drinking water on
the Susquehanna River. (7)

HEW discontinued the 24-
hour operation of its Command Post in Rockville. (6)

810-1845

Plume flight measurements
indicated a reading of 0.05
mr/hr at 500 feet altitude,
one mile from the plant. (1)

2055-2125

Plume flight indicated a
level of 0.05 mr/hr at 500
feet altitude, one mile from
the plant. (1)

April 7, 1979
0600-0645

Plume flight measurements
indicated a reading of 0.04
mr/hr at 500 feet altitude,
one mile from the plant. (1)

0755

The DOE Command Post
reported that "all is quiet."
(4)

1000

The Evacuation Center at
Hershey Park Arena was
closed. (10-f)

1030-1100

Plume flight indicated a
level of 0.05 mr/hr at 500
feet altitude, one mile from
the plant. (1)

1234

The DOE/EOC sked Ray
Zintz in the DOE Strategic
and Contingency Planning
Division when the agency
could stop giving radio-
logical assistance according
to the Interagency Radio-
logical Assistance Plan
(IRAP). Zintz, who
aided in drawing up the
IRAP, believed that DOE
should remain as long as its
services were needed to
protect the public safety. (4)

1253		The DOE Command Post reported that 51 DOE and DOE contractor personnel were at the site. In addition, the DOE was making two flights a day from Washington to TMI transporting NRC personnel. (4)
1315		In an up-date, the DOE Command Post reported 56 DOE and DOE contractor personnel at TMI. (4)
1805-1845		Plume flight indicated 0.05 mr/hr at 500 feet altitude, one mile from the plant. (1)
1955	Metropolitan Edison began lowering reactor coolant system pressure in 50 psi increments, a step towards cold shutdown. (15)	
April 7, 1979 2255-2335		Special plume flight to monitor a plant release. Instruments showed a level of 0.2 mr/hr at 380 feet altitude, one mile form the plant. (1)
April 8, 1979 0600	Hydrogen concentration in containment at about 1.85%; degassing continued. (15)	
0620-0700		Plume flight indicated levels of 0.3 mr/hr at 650-700 feet altitude, one mile from the plant. (1)
0900-0945		Plume flight measured radiation levels at 0.03-.04 mr/hr at 500 feet altitude, one mile from plant. (1)
1033		At this time 47 DOE and DOE contractor personnel were at TMI. (4)
1320	Reactor coolant system began to heat up, due to a decrease in steam generator level. (15)	
1535-2015		During this period there was a continuing discussion between the DOE/EOC, the NRC, ORNL, Bettis, and SRL to handle the analysis of 55 gallon samples of primary core coolant from the reactor. All arrangements with the laboratories were completed by 2015. (4)
DURING THE DAY		Anti-nuclear rally at the Harrisburg State Capitol. (10-f)

1805-1830		Plume flight measured levels of 0.05 mr/hr at 500 feet altitude, one mile from the plant. (1)
April 9, 1979 MIDDAY		Denton, in a press conference with Thornburgh, said that the nuclear crisis was over. Thornburgh said that women and children within a 5 mile radius of TMI could return home. (10-e)
1238-1308		Plume flight indicated 1.5 to 2.0 mr/hr at 500 feet altitude, one mile from the plant. (1)
1805-1848		Plume flight in strong wind and rain indicated levels of 1 mr/hr at 500 feet altitude, one mile from the plant. (1)
April 10, 1979 AM	TMI emitted some radiation. More releases expected. (10-e)	
0600	As of this time, hydrogen concentration in containment about 1.7%; degassing continued. (15)	
0627-0800		Plume flight launched in response to an announced release. Flight found 0.1 mr/hr readings at 200 feet altitude, one mile from the plant. (1)
1500		The NRC requested an immediate analysis on the primary core coolant sample after delivery to Bettis. (4)
DURING THE DAY		President Carter chose John Kemeny, President of Dartmouth College, to head the President's Commission on TMI, (10-d)
		At the request of NRC, a whole body counter was set up in Middletown by PaDER to scan citizens. (15)
1833-1913		Plume flight found 0.015 mr/hr levels at 500 altitude, one mile from the plant. (1)
April 11, 1979 0917		Joe Deal discussed with the DOE/EOC the position regarding the duration of DOE support to the NRC and the State of Pennsylvania at TMI. (4)

0917-0952		Plume flight during which "extremely low activity" was monitored. (1)
DURING THE DAY	Degassing operations continued; hydrogen concentration in containment about 1.8%. (15)	Bureau of Radiological Health shifted from a 24-hour a day schedule to a 12-hour a day schedule. (16)
1700-1743		Plume which found no visible plume. (1)
1840		The DOE asked the NRC to coordinate all requests for analysis of the primary core coolant samples through the DOE Command Post at Capital City Airport. (4)
April 12, 1979 0115	Degassing operations now completed; hydrogen concention in containment about 1.6%. (15)	
0938-1016		Plume flight indicated 0.03 mr/hr at 500 feet altitude, one mile from the plant. (1)
1205		DOE and DOE contractor personnel at TMI stood at 44. (4)
DURING THE DAY		South Carolina blocked TMI's radioactive waste from entering the state. (10-e)
		Laboratory tests showed that an insignificant amount of uranium melted during the accident. (10-c)
1510-1603		In rain and fog a plume flight found no readings above background. (1)
April 13, 1979 0115	Hydrogen recombiner stopped running due to burned out heaters. (15)	
0908-0940		Plume flight indicated mostly background readings. (1)
1003	Cooldown of the primary coolant system initiated. (15)	
1115	By this time 292 local residents have been scanned with a whole body counter in Middletown. Scan results showed no radiation levels above normal body levels. (15)	

1130	The Command Post reported 49 DOE and DOE contractor employees at TMI. (4)	
1454-1533	A special plume flight was launched in conjunction with the opening of the shielded doors in the auxiliary building. While the peak reading was 1.03 mr/hr, only one set of the doors were opened. (1)	
DURING THE DAY		Penrose Hallowell cancelled his advisory for farmers to feed cattle stored grains. Since March 28, his office had been sampling milk and grass. He said that no I-131 had been found in any grass. (10-b)
April 13, 1979 DURING THE DAY		Jack Watson named EPA as the coordinator for eight federal agencies involved in radiation data collection at TMI. (7)
2145-2230 and 2252-2335	Two plume flights were flown consecutively to monitor possible releases during the sampling of the primary core coolant system. The first flight found essentially background levels of radiation, 0.003-.005 mr/hr; the second flight had a maximum reading of 0.007 mr/hr. (1)	
April 14, 1979 1138-1221	Plume flight indicated a radiation level of 0.012 mr/hr at 250 feet altitude, one quarter mile from the plant. (1)	
2152-2350	Plume flight indicated a reading of 0.03 at 300 feet altitude, one half mile from the plant. (1)	
April 18, 1979 1000	DOE Command Post at Capital City Airport was closed down. (4)	

INDEX

206